# Río-Hortega's Third Contribution to the Morphological Knowledge and Functional Interpretation of the Oligodendroglia

# Río-Hortega's Third Contribution to the Morphological Knowledge and Functional Interpretation of the Oligodendroglia

*José R. Iglesias-Rozas*
Eberhard Karl University Tübingen, Germany

*Manuel Garrosa*
Faculty of Medicine and Institute of Neurosciences of Castile and Leon (INCYL), University of Valladolid, Spain

AMSTERDAM • BOSTON • HEIDELBERG • LONDON • NEW YORK • OXFORD
PARIS • SAN DIEGO • SAN FRANCISCO • SINGAPORE • SYDNEY • TOKYO

Elsevier
32 Jamestown Road, London NW1 7BY, UK
225 Wyman Street, Waltham, MA 02451, USA

First edition 2013

Copyright © 2013 Elsevier Inc. All rights reserved

Previously published in the Spanish Language by Trab. Lab. Histol. Patol. 1928 and Memorias de la Real Sociedad Española de Historia Natural, 1928.

No part of this publication may be reproduced or transmitted in any form or by any means, electronic or mechanical, including photocopying, recording, or any information storage and retrieval system, without permission in writing from the publisher. Details on how to seek permission, further information about the Publisher's permissions policies and our arrangement with organizations such as the Copyright Clearance Center and the Copyright Licensing Agency, can be found at our website: www.elsevier.com/permissions

This book and the individual contributions contained in it are protected under copyright by the Publisher (other than as may be noted herein).

**Notices**
Knowledge and best practice in this field are constantly changing. As new research and experience broaden our understanding, changes in research methods, professional practices, or medical treatment may become necessary.

Practitioners and researchers must always rely on their own experience and knowledge in evaluating and using any information, methods, compounds, or experiments described herein. In using such information or methods they should be mindful of their own safety and the safety of others, including parties for whom they have a professional responsibility.

To the fullest extent of the law, neither the Publisher nor the authors, contributors, or editors, assume any liability for any injury and/or damage to persons or property as a matter of products liability, negligence or otherwise, or from any use or operation of any methods, products, instructions, or ideas contained in the material herein.

**British Library Cataloguing-in-Publication Data**
A catalogue record for this book is available from the British Library

**Library of Congress Cataloging-in-Publication Data**
A catalog record for this book is available from the Library of Congress

ISBN: 978-0-12-411617-7

For information on all Elsevier publications
visit our website at store.elsevier.com

This book has been manufactured using Print On Demand technology. Each copy is produced to order and is limited to black ink. The online version of this book will show color figures where appropriate.

**Cover painting by Dr. José Rafael Iglesias-Rozas**

Dedicated to our wives María Reyes and Sonata,
and to all those keen on neurohistology

# Contents

# Foreword

The outstanding importance of the scientific contributions made by Pío del Río-Hortega to the field of neurology has to be gauged from the fact that of the four cell types constituting nervous tissue—neurons, astrocytes, oligodendrocytes, and microglia—he discovered the last two.

The discovery of the microglia cells, sometimes called *Hortega's cells*, is now fully attributed to Río-Hortega, even though previous partial descriptions were made by other scientists such as William Ford Robertson. However, the discovery of the oligodendroglia by Río-Hortega, which is perhaps of greater importance than that of the microglia, has been frequently ignored, probably because his original papers describing the discovery were written and published only in Spanish.

The authors of this book, seeking to remedy such a shortcoming, have already published the English translation of the two original articles in which Río-Hortega first described the oligodendroglia; now, with the translation of his third article as presented here, they complete his outstanding written contributions in this field. This third article gives a detailed and comprehensive account of the various morphological characteristics of the oligodendroglia; the account is still of such value that has been improved only by the contemporary use of electron microscopy and developments in immunohistochemistry.

The translators had to deal with the particular problem of handling the style of scientific language used in Río-Hortega's time, which was of a more rhetorical and complex construction than today's. They have been successful, I believe, in remaining as close as possible to the original text while clearly conveying its meaning.

In 1928, the year this article was published, Juan Manuel Ortíz Picón, a disciple of Río-Hortega, joined his laboratory; years later he described it in his memoirs as a modest place, carpenter's square in shape, where about a dozen people worked, each one at a desk provided with a microscope and a series of bottles containing stains and common reagents. There, Río-Hortega imparted his masterly teaching, as Ortíz Picón has described: "He was a most skillful technician and inventor of original silver impregnation methods of nerve tissue. He taught his disciples to look through the microscope with an attentive and continued vision so as to gain the ability to analyze and interpret even the most minute details of histological images."

Perhaps this is the key to Río-Hortega's discoveries and the main reason why this new edition can be useful: to achieve a precise and detailed description of the morphological characteristics of glia cells, their types, locations, and relationships that can be used to construct new morphofunctional hypotheses.

The authors, with the present book, now make accessible to neurobiologists what may be regarded as the most important article of this great Spanish histologist.

Prof. Dr Manuel J. Gayoso
Head of the Department of Cell Biology,
Histology, and Pharmacology, Faculty of Medicine
Director of Río Hortega Museum
University of Valladolid, Spain
November 12, 2012

# Acknowledgments

The authors most sincerely thank the Río-Hortega family, and especially Dr Juan del Río-Hortega Bereciartu in whose ownership the papers of Pío del Río-Hortega are held. With great friendliness and generosity, they have facilitated the use of those papers and provided us with some of Río-Hortega's own illustrations. The authors also gratefully acknowledge the assistance of Prof. Manuel J. Gayoso, director of the Río-Hortega museum in the University of Valladolid, for providing the original article and for his wise comments and criticisms. We are indebted to our colleagues Teresa Iglesias-Rozas, Brendan Purcell, and Cliona McCovern for their time and patience in reading the draft and for the valuable suggestions they have made for its improvement. We also wish to thank the Cajal Institute in Madrid and its librarian, Mrs. María Ángeles Langa, for the use of institute's facilities and help in obtaining illustrations from the original article.

Pío del Río Hortega

# Pío del Río-Hortega (1882–1945): A Biographical Sketch

Pío del Río Hortega was born on May 5, 1882, in Portillo, a small village near Valladolid, Spain. He was the fourth of eight brothers and sisters in a family of landowners who decided to move to the city so their children could receive good schooling and higher education. In Valladolid, the young Pío attended secondary school and took drawing courses in the School of Fine Arts, a skill that was to serve him throughout his life. In 1899, at age 17, he started medical school at the University of Valladolid, an institution famous for its historical tradition; its Faculty of Medicine is considered the oldest medical school in Spain, with documents proving it has existed for more than 600 years. The study of histology under the guidance of Leopoldo López García, professor of histology and anatomic pathology, deeply impressed Río-Hortega. His keen interest for this subject together with his excellent practical skills led the professor to appoint Río-Hortega as his honorary assistant. Soon afterward he got the position of tenured intern student of anatomy, Salvino Sierra and Leonardo de la Peña being his tutors. Under this status, he published his first paper in 1903, in *Boletín del Ateneo de Alumnos Internos* (*Bulletin of the Athenaeum of Intern Students*).

After the prescribed 6 years of medical studies, Río-Hortega received his degree in 1905. He worked as a general practitioner in his native village between 1908 and 1910, although he lived with his older sister María in the nearby historical and larger town of Olmedo. During this time, he also devoted himself to preparing for a doctorate in histology, which he was awarded in 1909 by the Central University of Madrid for the doctoral thesis he presented titled *Etiology and Pathologic Anatomy of Brain Tumors*.

In 1912, he moved to Madrid seeking to work with Nobel Prize winner Ramón y Cajal, but the already famous professor showed no interest in him. He succeeded in staying for a short time with F. Tello, one of Cajal's favorite students; he then joined another of Cajal's disciples, Nicolás Achúcarro (1880–1918), whose laboratory was located in the Museum of Natural Sciences, far from Cajal's facilities. In this laboratory, Río-Hortega became acquainted with the field of investigation into normal and pathological neuroglia; he learned methods of metal impregnation, mainly Achúcarro's tannin-ammoniacal silver, Cajal's formol-uranium and gold sublimate, Bielschowsky's method, and so on; and became captivated by the microscopic images. Based on their complementary traits of character and scientific interests, deep bonds of friendship soon developed between Río-Hortega and his mentor, N. Achúcarro, who was only a couple of years older.

In 1913, Río-Hortega obtained a competitive scholarship to study abroad and went to Paris to learn histological techniques with A. Prenant and pathological anatomy with M. Letulle. In 1914, he moved to Berlin to study at the Koch Institute with Joseph Koch but was forced to return to Spain at the outbreak of the First World War. Back in Madrid, Achúcarro's laboratory had been transferred to the same building as Cajal's Laboratory of Biological Investigations in the former Velasco Museum on Atocha Street. In 1915, Río-Hortega spent 3 months in London at the Imperial Cancer Research Foundation with J. Murray and Middlesex Hospital with L. Barlow.

In Achúcarro's new laboratory, scientific production grew quickly. Between 1913 and 1917, Río-Hortega published 27 papers on various histological topics, including the structure of the ovary, smooth muscle tissue, the fine texture of cancer cells and their epitheliofibrils, Paneth's cells of the vermiform appendix, the centrosome of neurons and neuroglial cells, and gliosomes and gliofibrils. He started to use his own methods while modifying Achúcarro's tannin-ammoniacal silver method. The relevance of these papers attracted Cajal's attention, even though some members of Cajal's school did not approve. By then, both Cajal and Achúcarro had studied the cytology of neurons and astrocytes, although the nature of the cells that Cajal called *apolar corpuscles* or *the third element* remained unclear.

Achúcarro became seriously ill with Hodgkin's lymphoma. As a consequence, Río-Hortega became interim director of the laboratory. He was then appointed secretary of the Spanish Society of Biology, whose president was Cajal. He was also nominated as interim assistant of Cajal's chair. In 1918, Río-Hortega attained a great breakthrough with his invention of a new staining method with ammoniacal silver carbonate that permitted him, in 1919, to determine what this "third element" was and identify it with a new cell type: *microglia*. This discovery placed Río-Hortega in a delicate position with Cajal because the latter held different theories about the third element and had criticized Río-Hortega's findings.

Then a new appointment was taken by Río-Hortega, namely, that of assistant to the laboratory of the Provincial Hospital of Madrid. He was also awarded in 1919 the first Achúcarro Prize, which had been established in remembrance of his colleague, who passed away the year before. Only after Achúcarro's death did Río-Hortega realize that the stipend he was receiving was not given by the institution but directly by Achúcarro, who wanted to share his salary with him. He then traveled to Paris to the International Congress of Physiology to present his work, which was beginning to have international impact.

The relationship with Cajal became increasingly cool, and the death of Río-Hortega's mentor left him without valuable support. Together with misunderstandings and intrigues from some of Cajal's coworkers and, in particular, a porter in the institute, this lack of support led Cajal to expel Río-Hortega from the laboratory in 1921. This event deeply affected Río-Hortega and he became ill. He had always wanted to work with his venerated idol, despite the fact that Cajal never showed him even one slide. Río-Hortega was in bed for several days with a high fever and was looked after by his inseparable friend and guesthouse mate, Nicolás Gómez del Moral. In the meantime, Cajal, who knew Hortega's outstanding scientific value, took the necessary steps to procure a new place for his laboratory in La Residencia de Estudiantes of the National Board for Advanced Studies, a place where scientists such as J. Negrín, L. Calandre, G. Rodríguez Lafora, and P. Suárez were also working.

Río-Hortega continued his work studying the mesodermal origin of microglia, as well as their phagocytic functions in different pathological processes. In 1921, the silver carbonate method also permitted him to identify a new cell type of neuroglia that exhibited delicate cytoplasm extensions; he named it *oligodendroglia*. This led him to the hypothesis in 1922 that oligodendroglia were homologous to Schwann cells. He showed these papers to Cajal, who became more friendly with him. That same year they had a meeting in a café, where they reconciled. During their conversation, Cajal admitted that he was wrong about the third element and said he would rectify his position in forthcoming publications, giving due acknowledgment to the primacy of the discoveries of microglia and oligodendroglia to Río-Hortega. This promise was fulfilled in 1925.

By 1923, Río-Hortega had become internationally known and attracted several foreign scientists such as C. Da Fano, P. Bailey, W. G. Penfield, J. Turchini, A. Meyer, and H. Cushing, who with Bailey offered him a laboratory in the Peter Bent Brigham Hospital in Boston. Meyer also offered a laboratory in The Johns Hopkins Medical School in Baltimore. Río-Hortega, however, loved Spain and decided to remain in Madrid. During this time, he worked on the pineal gland and with Penfield confirmed the existence of microglia in gliomas and in the cerebral cicatrix processes in epilepsy. Although 1 year earlier A. Gans had used the term "Río-Hortega's third element," German scientists A. Metz and H. Spatz published an important paper on microglia in 1924 in which they supported Río-Hortega's theories and named these cells *Hortega's cells* in his honor.

In 1925, Río-Hortega gave seminars on the histology and histopathology of the nervous system at the University of Montpellier, to the Society of Biology of Paris, and in different cities in Argentina and Uruguay. He refused again a profitable position as neuropathologist offered to him by the John Rhea Barton Hospital of Philadelphia. In 1926, he was appointed president of the Spanish Royal Society of Natural History. He presented his investigations on the pineal gland at the annual meeting of the Anatomical Society celebrated in Liège. He was then appointed honorary professor of the University of Valladolid and 1 year later member of the Society of Biology of Paris. More seminars were given in Madrid, Barcelona, Paris, and Budapest while he also began working on nervous system neoplasms.

In 1928, Río-Hortega was appointed head of the Laboratory of Experimental Oncology of the Prince of Asturias National Institute of Oncology, where he spent time working happily and steadily while surrounded by his disciples, among them Isaac Costero, M. López Enríquez, C. Collado, F. Jiménez de Asúa, R. Vara López, R. Alberca, A. Llombart Rodríguez, and J. M. Ortíz Picón. He was nominated as a candidate for the Nobel Prize for 1929. During this same year, he lectured in France, Mexico, and Cuba and was appointed director of the Laboratory of Pathologic Anatomy and Oncology at the brand new Hospital of Valdecilla (Santander, Spain), where he gave seminars every summer. In 1930, one of his speeches on the reform of university medical education, given at the Faculty of Medicine of Madrid, upset the most conservative sectors in the university and the National Academy of Medicine, where some of his old enemies occupied positions.

Río-Hortega published more papers on meningoexotheliomas, meningeal endotheliomas, and gliomas. Brain tumors were sent to him for diagnosis from different

American and French hospitals and later also from Canada and England. More foreign scientists came to stay in his laboratory, including L. Van Bogaert, J. B. Gaylor, E. Moniz, H. G. Creutzfeldt, and H. Urban. In 1931, his increased scientific reputation led him to Germany on the invitation of professors M. Borst, W. Spielmeyer and H. Spatz at the University of Munich, C. Benda and A. Goldscheider in Berlin, L. Brauer and O. Vogt in Hamburg, E. Kallius in Heidelberg, and L. Aschoff in Friburg—even though his publications had only appeared in Spanish. This trip to Germany was an enormous success and was followed by a trip to France and the Institute of Oncology of Paris on the invitation of G. Roussy.

A great political change came to Spain in 1931 when the monarchy ended and a republic was proclaimed. This same year, Río-Hortega was appointed by Minister A. Lerroux as a member of the Council for Cultural Relations, together with other Spanish figures such as R. Menéndez Pidal, G. Pittaluga, G. Marañón, and G. Rodríguez Lafora. Río-Hortega published an article in the prestigious newspaper *El Sol* in which he criticized the university institution and its stagnation, creating more enemies among Spanish academics. They made their influence felt when some faculties of medicine, some National Academy members, the Federation of University Students, and several provincial governments sought the establishment of a personal chair on "Normal and Pathologic Histology of the Nervous System" for Río-Hortega—and no answer was given. He accepted a new distinction as member of the Committee of the Board of Museums of Sciences. In 1932, he was appointed director of the National Institute of Cancer and 1 year later also became councilor of the National Council of Health. Once again he was nominated for the Nobel Prize, but also once again he did not receive the necessary support from his own country.

Río-Hortega's work on the microscopic anatomy of the tumors of the central and peripheral nervous system was presented in the International Congress of Scientific and Social Struggle against Cancer in Madrid. This work has become a milestone in its field. In 1934, he was nominated member of different scientific societies in the Americas and Europe. However, in his own country, he continued to be a victim of public accusations. In addition, he was not given, as it was expected, the vacant chair left by Cajal in the National Academy of Medicine after his death that year; one of Río-Hortega's own disciples was appointed instead. This issue caused G. Marañón and G. Rodríguez Lafora resign from the academy. These and other intellectuals of his time offered him homage, including Valle Inclán, García Lorca, and Jiménez de Asúa, who openly spoke in his defense. In 1935, he was nominated honorary fellow of the New York Academy of Sciences and of the National Geographic Society of the United States.

In 1936, the Spanish Civil War broke out. Río-Hortega spent a few months in Paris before returning to the Institute of Oncology in Spain, which was located on the university campus and just on the battlefront. After 2 weeks under bombardment, he decided to send the institute's microscopes to Valencia, where the capital of the republic had been moved. His friend Nicolás Gómez and niece Asunción Amo tried to save as many slides as they could by putting them in shoe boxes while the bombs were falling. One explosion caused Asunción to faint, but she and Gómez managed to return home safely in an ambulance. At this juncture, Río-Hortega also closed his laboratory at La Residencia de Estudiantes.

In 1937, Río-Hortega moved to Paris accompanied by his sister Felisa and niece Asunción. Nicolás Gómez del Moral joined them later. There he worked in the laboratory of histopathology of the Hospital of La Pitié supported by a Spanish government stipend to continue the research in order to complete the chapter on tumors of the nervous system in the *Diagnostic Atlas of Tumors* for the International League Against Cancer. Then he received the invitation of Professor H. Cairns to work in the Nuffield Institute in Oxford, where a duplicate of his laboratory in La Residencia de Estudiantes was built for him. He accepted, and the four traveled to Oxford, where he met other expatriate Spaniards such as J. Trueta, S. Ochoa, A. Jiménez Fraud, I. Costero, J. de Castillejo, Paulino Suárez, Salvador de Madariaga, and A. Duperier. As a patriot, he felt homesick for Spain, his native Castile, and his circle in the café, where he frequently gathered with other compatriots. These acquaintances praised Río-Hortega's refined warmth and admirable skill in drawing cells. In 1939, the government accepted accusations that he was "a republican" and "a Mason," and all positions and responsibilities he had won or been granted were withdrawn.

The University of Oxford appointed him lecturer and awarded him the title of Doctor Honoris Causa in 1939, and Sherrington being his academic supporter and patron. He was also admitted as honorary fellow of the Senior Common Room of Oxford Trinity College. The Second World War forced him to move again and accept the invitation of the Spanish Cultural Institution in Buenos Aires, Argentina. He was established there in 1940 and began to work in the Hospital of Santa Lucía in 1941 while a new duplicate of his laboratory of La Residencia de Estudiantes was being built for him. This new laboratory was named *Ramón y Cajal Laboratory* by Río-Hortega himself, who felt happy working there and soon created a school with disciples such as M. Polak, who had worked with him in Oxford, J. Prado, L. Zimman, and J. Thenon. Río-Hortega became Doctor Honoris Causa and Extraordinary Professor of the University of La Plata and crossed the border several times to give courses in Montevideo. In 1942, he edited the journal *Archivos de Histología Normal y Patológica*, where he published his new papers on tumors of the nervous system, staining methods, glia of ganglion neurons, and so on and established the concepts of *angiogliona* and *neurogliona* as elemental units of glia.

The biopsy of a nodule appearing in his own urethra was diagnosed by him as malignant in 1944. He participated in Montevideo in an event to pay homage to Cajal and commemorate the 10th anniversary of Cajal's death. A few more papers appeared in 1945, but he was admitted in serious ill health to the hospital of his friend, Dr Avelino Gutiérrez, where he died on June 1 surrounded by his friends and disciples. Representatives of the Spanish, British, French, and American worlds of science spoke at the time of his funeral, as did personal friends and representatives of the Argentinian, French, Spanish, and British intelligentsia. No official representation was sent by the Spanish government.

Pío del Río-Hortega made invaluable contributions to histology and neuropathology that placed him in an outstanding position in the history of science. No doubt he deserved the Nobel Prize, but as also happened to some others of his stature, extrascientific reasons intervened to prevent him from receiving such a distinction. The glory and brilliance reached by Río-Hortega in science, to which he devoted his life, was echoed by international recognition and awards throughout Europe and the Americas, as well as by the chain of outstanding disciples who continued his work and taught in those countries and institutions where the different wars had sent him. He always wanted to return to the Institute of Oncology in Madrid, but destiny did not allow his wish by cutting his life short at 63 just when the maturity of his life might have yielded more abundant scientific production. Forty years after Rio-Hortega's death, a movement advocating due recognition of his life and scientific merit emerged in his native country, leading to his desired return, albeit posthumously, in 1986. His mortal remains were placed in the Pantheon of Illustrious Men in Valladolid.

# Preface

Río-Hortega first described the microglia in an article published in 1919 where he identified other cells with very fine processes that he named "interfascicular glia". Two years later, he accurately characterized a new type of neuroglia that could not be considered protoplasmic or fibrous astroglial cells and named them "glia with very few processes" or "oligodendroglia" (1). He observed that these cells showed only a scant number of ramifications (oligo = few; dendro = tree). These types of cells included the "interfascicular glia", but this name was not suitable because astrocytes may be interfascicular too and oligodendrocytes are also present in other locations than fascicles. In a second pioneering article published one year later, he also put forward the possibility that these newly identified oligodendroglial cells could be homologous with Schwann cells in the central nervous system (2). We translated these two seminal articles on oligodendroglia into English for the first time and published in the journal *Clinical Neuropathology* in 2012 (4)(5).

A third article by Río-Hortega on oligodendroglia appeared six years later titled "Tercera aportación al conocimiento morfológico e interpretación functional de la oligodendroglía" (*Third contribution to the morphological knowledge and functional interpretation of the oligodendroglia*) (3) that is the one which we edited in English in this book. This work from 1928, dedicated to Dr. Avelino Gutiérrez, is an extensive paper that includes a description of the different types of oligodendroglia, their presence and distribution throughout the cerebral cortex, white matter, cerebellum and spinal cord, also their ontogeny, functional homology with Schwann cells of the nerves as well as their pathological and postmortem changes. Río-Hortega claimed that "in fact only the microglia could be regarded as the third element"; neurons were the first element and the neuroglial cells the second.

Since then, the name oligodendroglia replaced the terms "adendritic elements" "apolar elements", "naked nuclei", "round cells" or "cuboidal cells" that were poorly characterized and not standardized between different authors. This terminological confusion allied to the fact that Río-Hortega's publications originally appeared in Spanish only and therefore were not easily available to the international community. This could account for the lack of wider recognition of the discovery of oligodendroglia by Río-Hortega in non-Spanish speaking countries. In 1900, Robertson described interstitial cells in the nervous tissue with his platinum method that he classified as 'mesoglia cells' supposing that they were of mesodermal origin. His descriptions, however, did not determine whether these mesoglia cells corresponded to microglial or to small oligodendroglial elements. Similarly, Cajal using his gold sublimate and formol-uranium methods was not able to identify oligodendroglial cells either. This made him criticize Río-Hortega's discoveries and led to an issue about the third element between them. A few years later, however, Cajal accepted that Río-Hortega was right and gave him the recognition of having been the first to make these discoveries.

The reading of classic works in neurohistology, such as this third contribution on oligodendroglia by Río-Hortega, widens our perspectives by presenting fundamental concepts and ideas. We will also come to know the concrete details on which these concepts and ideas depend and will learn about those qualities that make Pío del Río-Hortega a person of great scientific stature, who deserved to be twice nominated for the Nobel Prize. Unfortunately, it was not be awarded to him due in part to the interference of some colleagues and political interests among those who should have supported him in his own country.

There are some substantial discrepancies between Río-Hortega's descriptions of the oligodendroglia, their progenitors and in particular his classification of four types, and the immunological phenotypes of modern understanding that tended to ignore the morphological facts and distinctions altogether. Indeed, it is very rare to find the proper attribution due to Río-Hortega's work and achievements in modern literature, even in very recent studies and reviews concerning oligodendroglia. For example, whereas Hortega claims that the most suitable study on oligodendroglia should be performed in large animals such as the monkey, cat, dog, sheep, and donkey, nowadays most of the studies are conducted in smaller animals like mice, rats or rabbits. Also, he has shown that not all oligodendroglia develop myelin, since 1st type (formerly called "of Robertson") have perineuronal functions without forming myelin. According to Río-Hortega, whether oligodendroglia develops myelin or not they should be considered cells of the same species, although they may show phenotypic variations in differentiation or function. These points were made over 80 years ago, and surprisingly enough they are neither mentioned nor clarified or developed in modern immunological works.

Río-Hortega stated that oligodendrocytes sometimes may have many extensions, as it is the case with protoplasmic and fibrous astrocytes. He had already described, particularly during the period of intense myelinization in young animals, those cells that could be considered intermediates between astrocytes and oligodendrocytes, as they are also described now with immunological methods. Penfield, and subsequently López Enríquez, fully corroborated these findings acknowledging that there are no transformations between astrocytes and oligodendrocytes in pathological cases. Río-Hortega had clearly stated: "The only difference between the two neuroglial elements is the lack of vascular insertions of oligodendrocyte". The differences between origin, morphology and function of the microglia and oligodendroglia are thoroughly and accurately treated in the present work.

Throughout the text, the reader can appreciate Río-Hortega's remarkable and extraordinary capacity and technical ability in accomplishing the procedure of staining tissue elements of the central and peripheral nervous system. Coupled with this is his outstanding ability, first to observe and then to interpret the function of the isolated cells and their interactions with different neighboring tissue elements with the help of microscopes of his time.

During the years of our histological education and up to the present, neurohistological works have not been published on oligodendroglia, microglia and astroglia cells that rely on Río-Hortega's techniques either as a primary base and/or in combination with inmunohistochemical techniques. The diversity of reasons for this neglect is complex and varied, but we believe that the most important one has been the lack of a translation of Hortega's work into English or German. Moreover, the histological "silver techniques" for oligodendroglia, as easy or difficult as immunological techniques currently are, in our opinion, have been *inexplicably ignored* by neuropathologists and pathologists for years. This can be seen, for example, in the classical book "WHO Classification of Tumours of the Central Nervous System" (IARC, Lyon 2007), especially in the chapters concerning oligodendrogliomas and oligo-astrocytomas. In addition, the specific and relevant comparisons between the "old" techniques of silver impregnations from Río-Hortega and the modern techniques of immunology are missing from present discussions. Our experience has been that the monoclonal antibody MAP-2 (Anti-Microtubule-Associated Protein (2-SIGMA) shows the best results in human tissue *(Contact:jiglerozas@t-online.de)*.

The new generations of neuroscientists who are not familiar with the "silver techniques", could definitively find this book as an invaluable stimulus. It will teach them first to *learn to observe* and then *to interpret faithfully* the observations gained using current, modern genetic and immunological techniques with the help of excellent modern microscopes. We believe it is scientifically inappropriate to create new genetic and immunohistological classifications of oligodendroglial cells without having read, or referred to, Río-Hortega's original articles. Furthermore, oligodendroglial cells are involved in many diseases of the central nervous system with dramatic impact on the patient, such as: demyelinating diseases, leukodystrophias, bipolar disorder, schizophrenia and cerebral or spinal injuries. Thus, the precise study, the understanding of the pathophysiology, the management of oligodendroglial cells and their progenitors is very important for shedding new light on the treatment of these diseases.

In editing this book, the translators make the first English version of Río-Hortega's "*Third contribution to the morphological knowledge and functional interpretation of the oligodendroglia*" available to the scientific community. We have translated the origin text as literally as possible trying to respect the original style, although Río-Hortega's highly rethorical syntax could make readability difficult in many instances. Only where absolutely necessary, slight modifications have been made. All illustrations were taken from the original publication and are reproduced, in the majority of cases, without alteration. Apart from corrections of dates and obvious typographical errors, we have allowed the textual citations to stand in the form in which they originally appeared.

As mentioned above, we are convinced that this book may be helpful not only for its historical value but also for its relevance to current research and neuropathological diagnostics. We also carried out this work for personal reasons that go back to our academic education when we were students of Prof. Luis Zamorano, who had learned the techniques directly from Río-Hortega at "La Residencia de Estudiantes" in Madrid. In addition, Dr. Iglesias-Rozas improved his knowledge of silver impregnation methods when working at the Institute of Neuropathology (Prof. Dr. J. Cervós-Navarro, Director) of the Free University of Berlin together with Prof. Moisés Polak, a direct disciple of Río-Hortega. Finally, this book has the purpose of paying a modest tribute to this great Spanish histologist and histopathologist.

## Preface references:

(1) P. del Río-Hortega: Estudios sobre neuroglía. La glía de escasas radiaciones (Oligodendroglía), Bol. de la Real Soc. Esp. Hist. Nat. 21:63–92, 1921, also published in P. del Río Hortega. Estudios sobre Neuroglía. La glía de escasas radiaciones (Oligodendroglía). Trab. Lab. Histol. Patol. X:1–30, 1921. and in Arch. Neurobiología, 2:16–43, 1921.

(2) P. del Río-Hortega: ¿Son homologables la glía de escasas radiaciones y la célula de Schwann? Bol. de la Soc. Esp. de Biol. X(I):25–28, 1922, and P. del Río-Hortega: ¿Son homologables la glía de escasas radiaciones y la célula de Schwann? Trab. Lab. Histol. Patol. 26:1–4, 1922, and Río-Hortega, P. del: ¿La glia a radiations peu nombreuses et la cellule de Schwann sont-elles homologables?. Comp. Rend. Her. Domad. Soc. Biol. 2:818–820, 1924. (*Translation into french*).

(3) P. del Río-Hortega: Tercera aportación al conocimiento morfológico e interpretación funcional de la oligodendroglía. Memorias de la Real Sociedad Española de Historia Natural, 14:5–122, 1928, and P. del Río-Hortega: Tercera aportación al conocimiento morfológico e interpretación funcional de la oligodendroglía. Trab. Lab. Histol. Patol. Vol 8-100, Tomo XIV:5–122, 1928.

(4) P. del Río-Hortega. (Translated and edited from the original Spanish (1921) by José R. Iglesias-Rozas and Manuel Garrosa): Studies on neuroglia. Glia with very few processes (oligodendroglia) by Pío del Río-Hortega. Clinical Neuropathology, 31(6):440–459, 2012.

(5) P. del Río-Hortega (Translated and edited from the original Spanish (1922) by José R. Iglesias-Rozas and Manuel Garrosa): Are the glia with very few processes homologous with Schwann cells? by Pío del Río-Hortega. Clinical Neuropathology, 31(6):460–462, 2012.

# 1 Introduction

If we had to take risks now for the first time in the difficult task of raising awareness of the glia with very few processes or oligodendroglia, we would not hesitate to support our general description of 1921 to a large extent. Using fragmentary observations, often achieved with technically inadequate methods, we then strove to build a theoretical structure to demonstrate the existence of an important yet largely ignored variety of neuroglial elements. Although with barely firm foundations, this structure apparently had sufficient solidity and we believed would become perennial simply by improving morphological details.

From 1921 to 1928, we did not desist in our efforts to discern and improve oligodendroglia architecture, and by tackling it yet again, we are convinced (as certain as what we have claimed with regard to microglia) our fundamental concept and supporting details will not be seriously contested.

Despite having predicted significant surprises when researching further into the new variety of neuroglial cells, we were only strongly impressed by the newly discovered details regarding the morphological and functional relationship of oligodendroglia and Schwann cells.

Conjecturing as to the similarity of trophotonic functions—that is, isolation, nutrition, and support for both categories of elements in relation to nerve fibers, we questioned in 1922, we put forward a hypothesis as an answer: Are the glia with very few processes homologous with Schwann cells? Today, with sufficient arguments in our power, we do not hesitate to answer this question as follows: *The oligodendroglia in encephalo-medullar centers is absolutely homologizable with the Schwann cells of the nerves.*

Thus, Cajal's hypothesis (1913) regarding "compensation or replacement" of the "apolar elements" of the brain and Schwann corpuscles is thus confirmed.

However, as stated in our previous research documented, the so-called *apolar corpuscles* are identified with the glia with few radiations to which Cajal's hypothesis certainly belongs.

There are no adendritic cellular elements in the complex interlocking of the normal central nervous system, where neurons, neuroglia (with polymorphisms), and microglia (first, second, and third elements, respectively) emit branched radiations. Nevertheless, there are still some erudites entrenched in tradition (partly his work) and supported by a multitude of research works, who are reluctant to exchange yesterday's truth with that of today.

In the group of the second element, since classical times, researchers noted, utmost importance was given to elements readily available for observation, that is, those evidenced by the Golgi method and defined with the expressions of protoplasmic glia or glia of short radiations and fibrous glia or glia of long appendages.

**Río-Hortega's Third Contribution to the Morphological Knowledge and Functional Interpretation of the Oligodendroglia.**
**DOI: http://dx.doi.org/10.1016/B978-0-12-411617-7.00001-8**
© 2013 Elsevier Inc. All rights reserved.

It took a long time, however, to determine the existence of apolar elements outside these morphological definitions. When Cajal had the brilliant insight that a third element existed and was distributed in the nerve centers, only a few researchers were clearly interested in knowing the corpuscles that all techniques, with remarkable consistency, showed up as almost protoplasm-free isolated nuclei and spherical silhouette barely altered by short dentils. Such meager corpuscles could not produce aesthetic emotions. And yet, these refractory cells for all protoplasmic staining, by their infinite number, for dissemination in encephalic areas and strategic location, next to neurons and nerve fibers, should have caught the attention of neurologists at one point and placed at a level comparable to classic neuroglia.

This moment has arrived for the so-called apolar elements, whose new physiognomy justified the denomination of glia with few radiations, in the foreground of neuroglial studies.

Cajal's criterion regarding the third element of nerve centers started to be unanimously accepted (if it had ever been taken into account) casting Robertson's very shallow and confusing study into oblivion, regarding mesoglial cells, where our research, inspired at all times by those of our masters, led to differentiation of microglia (mesenchymal) and oligodendroglia (ectodermal) among Cajal's apolar corpuscles.

At the time the pathological types of microglia were known, which absolutely ignoring everything related to normal forms and even its existence as a constant element of nervous tissue. As to the oligodendroglia, whose circumstantial morphological description and genuinely neuroglial interpretation which we were the first to perform, a descriptive outline in Robertson's publication already existed.[1]

The cytological details collected until 1922, explicitly showed, that the nerve centers—in addition to the known neuroglial variety—a system of gliocytes with typical features, whose main characters lay in the cell body (small, rounded, and polyhedral) and in the irradiated expansions of its contour (few, long, varicose, briefly branched, and tending to sheath the nerve tubes).

## Denomination

In our initial work on the glia with very few processes, while unaware of Robertson's discovery about one type of neuroglial element with two, three, or more short extensions, called mesoglial cells, a meaningful term had to be found, which was none other than *oligodendroglia*, which fit the reality of that moment precisely, *comprising the interfascicular glia together with the neuronal and vascular satellites.*

However, today the reality after 7 years of constant pursuit, is that not all cells under the term *oligodendroglia* should be considered Robertson's cells or even oligodendroglia,

[1] As there are still some authors, although only a few, who are doggedly searching literature prior to these studies, please forgive us for tackling this subject by a simply copying a few words by Percival Bailey: "... that del Río-Hortega in particular isolated and established the presence in the interstitial tissue of the brain of two cellular categories hitherto only vaguely suspected, which he called the microglia and the oligodendroglia."

a term proposed by us, since oligodendrocytes have been proven to have a greater number of processes than their term implies, although we hasten to say this is always below that emitted by protoplasmic and fibrous astrocytes, so the term remains relative.

The term *oligodendroglia* was so quickly accepted by neurologists, especially German- and English-speaking ones, manifested that a specific term was deemed necessary given the variety of elements clearly bearing neuroglial characters, scattered in the nerve centers and for which a vast variety of terms have encountered accommodation and apparent justification. Thus, the term *oligodendroglia* replaced *adendritic* or *apolar elements, bare nuclei, round cells, cuboid cells*, and so on, in histological descriptions.

Coining a term lacks scientific scope except when constituting a definition, as it tacitly accepts the concept motivating the same. Therefore, in the case of *oligodendroglia*, use of the term without discrimination or exceptions implies acceptance of the morphological concept we support. We do not consider the authors' decision to routinely use the term referred without value since it was clearly not preceded by rigorous testing of the corresponding objective data. Technical difficulties made it extremely difficult to confirm the accuracy of our descriptions. So on recalling the fear that led to the statement of difficult to verify facts, it was heartwarming to have authors encourage us, openly expressing their sympathy regarding the scientific issue of oligodendroglia. It is noteworthy that this fact causes us emotive interest, because against our ideas on oligodendroglia we had the most prestigious of our mentor regarding the apolar elements, and of those who like us oriented their behavior toward discipline and scientific rigor.

We spoke a moment ago about the value of the term oligodendroglia as opposed to that of polidendroglia. In fact, when one sees the number and division of branches sprouting from oligodendroglial elements, the term loses its strength, thus to reestablish its accuracy one must verify that the number of expansions irradiated from common astrocytes is infinitely superior. However, we would never dare state whether the branching of these are more detailed than the less copious processes of oligodendrocytes.

Since this term lacks accuracy or descriptive power, should it be replaced by another? We think not. Wherever possible, technical terms in histology should not follow a simple morphological plan but a physiological one. But how to create a physiological term when the function inspiring this is unknown? So let us leave the interim anatomical names *astroglia* (protoplasmic and fibrous macroglia), *oligodendroglia*, etc., as they stand until they can be justly replaced by *trophoglia, tonoglia, crinoglia*, etc., or others in line with this rigorously tested role, as opposed to the hypothetical one although based on abundant evidence.[2]

[2]It is not the first time we have felt the need to call some authors to attention, who feel the desire to change the nomenclature based solely on his/her personal whim. When a researcher runs out of patience when studying an already exhausted subject using certain techniques, he/she is unwilling to let his/her effort go unpublished drafting the corresponding article where the absence of objective novelties encourages him/her to create a classification or introduce a new word. Unfortunately, this is undoubtedly one of the mechanisms of scientific development in medicine and biology, where renewal is often (a question of fashion) external and not internal. Dressing up old ideas (i.e., renaming them), thus, modernity is achieved at little cost. As to the visibility, sought by the author, this is often achieved from among those for whom being up-to-date means latching onto strange names as opposed to acquiring new ideas.

## Concept

We have no doubt that the glia as a whole play a common role of protection and isolation, whereby according to their morphological types, are able to isolate and fill (protoplasmic varieties) or tone and support (fibrous varieties). It is also clear, however, that in addition to this passive/mechanical "presence" role, the neuroglia exert another nobler role with a more direct influence on the nervous elements by modifying their trophism, providing specific elements, neutralizing toxins, and so on. In this regard, we should not insist on the plausible hypothesis of the secreting neuroendocrine role, of the neuroglia.[3]

Oligodendroglia are fully understood in these activities given that their expansions form complicated plexus that unsheathe (isolating and protecting) cells and nerve fibers and create Schwannoid supportive arrangements around the medullated tubes. Their protoplasm produces a specific product which probably influences the trophic function of nerve fibers and elaboration of myelin.

This physiological concept of oligodrendroglia is combined with the current morphological concept summarized as follows: association of highly abundant neuroglial elements with generally small soma from which a variable number of filiform or laminar expansions arise and divide several times, follow the course of myelinated fibers (unsheathing them almost entirely) forming altogether a retiform plexus surrounding the entire nerve arrangements.

[3]See P. del Río-Hortega. Condrioma y granulaciones específicas de las células neurógicas. *Bol. de la R. Soc. Esp. de Hist. Nat.* 1925.

# 2 Bibliographic Background

If we were to recall here the extensive literature concerning the apparently neuroglial adendritic elements, we would require many pages and our information would ultimately be meaningless because their argument is not questionable: the identification of oligodendroglia with all sorts of nonstar-shaped corpuscles with the classic methods for neuroglia (except microglia).

But as a tribute to the researchers who put in the effort to unravel the mystery of the "adendritic" cells, we mention some important names. Some of their findings supported by evidence are now fully confirmed.

Among those who discussed the interfascicular situation of the "apolar" corpuscles are Jakob, Buscaino, Rosental, Eisath, Perusini, and Cajal.

The perineuronal clusters were studied—apart from those who investigated the neuronophagy—by Obersteiner, Golgi, Nissl, Andriezen, Lugaro, Alzheimer, and especially Cajal. In 1896, our master already held the view (but abandoned in 1914) that the cells with scant stellate cytoplasm normally forming pleiads, especially alongside the polymorph corpuscles of the cerebrum, had a neuroglial character.

As for the cell characteristics, just remember the proposed names of *bare nuclei* (Schaper), *indifferent cells* (Bonomo), *round cells* (Eisath), *cuboidal cells* (Cerletti), *preameboid cells* (Rosental), and *adendritic elements* (Cajal). All of these names have the descriptive importance of true definitions.

One of the most interesting and litigious points—the theoretical interpretation—has long been widely discussed. Three hypotheses scrambled to take prominent place:

1. That of Held, Alzheimer, Fieandt, Jakob, and Lugaro, which proposed that all the interstitial cells of nerve tissue that occur without expansions with neuroglial techniques are really branched.
2. That of Bevan-Lewis, Nissl, Robertson, Bonomo, Schaper, and Rosental, which proposed a kind of germ or undifferentiated gliocytes devoid of expansions.
3. That of Cerletti (accepted by Cajal) which held that the cuboid elements are cells with definitive characters and equipped with special function.

Clearly, the last is no longer a hypothesis because it has reached the category of fact. The first is now in fundamental agreement with the facts. Nothing seems to favor the idea that there are undifferentiated or germinal elements because oligodendroglia show essentially unchanging characters that correspond to a permanent functional adaptation.

Río-Hortega's Third Contribution to the Morphological Knowledge and Functional Interpretation of the Oligodendroglia.
DOI: http://dx.doi.org/10.1016/B978-0-12-411617-7.00002-X
© 2013 Elsevier Inc. All rights reserved.

## The "Mesoglia Cells" of Robertson

Previous work that was closest to the current reality for oligodendroglia is that of Robertson's memorable research, although it is far away from that. This neurologist in 1900 found in the cerebrum the existence of a new breed of neuroglial elements that, if confirmed to be of mesodermal origin, proved to be right the designation *mesoglia cells*.

The brief submission on behalf of Robertson by Dr. Clouston in the Neurological Society of Edinburgh said only the following:

> *The first fact to which I need direct the attention is that Dr. Ford Robertson has imagined a new method to examine the nervous tissue, depositing platinum on it. By using this platinum method he has shown, among other things, that what is called neuroglia is composed of two kinds of element instead of one, as is generally estimated. The neuroglia, as shown with this and other methods, is in contact with the arteries, fibers and cerebral cells, forming a general supporting medium. Dr. Robertson has discovered that in addition to this there is another type of glia, called mesoglial cells by him, consisting of a typical cell body, a nucleus and a number of processes. These extensions are not in connection with either the vascular substance or with the cells or nerve fibers. The mesoglial cells are completely different in appearance from the cells of neuroglia and are found in the white matter as gray matter, and in such abundance that Dr. Robertson found that there are as many mesoglia cells as neuroglia cells throughout the cerebrum. Sometimes they have no process, sometimes two processes, but the illustrations show a typical mesoglia cell in dog and man. The exact function of these mesoglia cells is not known for certain, but certainly not in any way they act as a support for the overall structure of the cerebrum. The mesoglia cells seem to have a phagocytic action in certain pathological states. They provide, if not completely, at least most of the amyloid bodies found in some chronic degenerations of the cerebrum.*

This descriptive outline by Robertson, not being accompanied by illustrations, lacked precision and showed only the belief of the wise Scottish scientist that neuroglial elements with small number of expansions existed in the cerebrum that were erroneously attributed to mesodermal lineage. No wonder then that Cerletti said that neither the morphological differences described nor different reactions with the platinum method allowed the creation of a new special category of interstitial cells or the assertion of a histogenetic difference for the two types of elements. In the book *Pathology in Mental Diseases*, Robertson insisted on describing his cells, accompanied this time by engravings that illustrated his observations better than the pure report of the morphological characters.

Regarding the form, the mesoglia cells range three to six delicate short processes in several directions, which are dichotomized up to three or four times, and then gradually decline their thickness toward their free ends and have small varicose dilations. The vast majority of cells are highly branched, although some only a little. Not all have processes; variations can be seen in the dog (which corresponds to the general description), the ram (which has wider cells with few and short appendages), and humans (who seem to have primarily slightly branched corpuscles).

Regarding functions, the mesoglia cells cannot serve as means of support because they lack vascular attachments and take no active part in the process of repair. Their role appears to correspond to the endothelial and connective cells; they are able to take in foreign particles (i.e., phagocytosis).

With respect to alterations, the mesoglia cells also resemble endothelial and connective elements. In irritative conditions such as general paralysis, they probably tend to proliferate, but they also lose their limbs becoming granular bodies as a result of phagocytic functions. The second change noted by Robertson referred to the formation of amyloid bodies.

In fact, because confirming the discovery by Robertson requires a long technique that is difficult to carry out, his ideas attracted few followers and remained almost unknown. Only a few authors speak about Robertson's mesoglia and do so without conviction. We vaguely remembered hearing something about mesoglia from our master Achúcarro, but we did not think it could have any relationship with our oligodendroglia. This means that when we called attention to this cell system, we ignored that many years before Robertson had seen morphologically similar elements, although functionally different regarding their interpretation from various points of view.

Indeed, although the forms of mesoglia cells correspond to the small types of oligodendroglia,[1] there are profound concept discrepancies between Robertson's ideas and ours.

Robertson is remarkably wrong

1. in describing the mesoglia cells as elements completely different from the neuroglia,
2. in claiming that they are not in connection with nerve fibers,
3. in maintaining that they do not act in any way such as support of the general structure of the cerebrum, and
4. in ascribing them phagocytic function.

We could say that Robertson had the very meritorious intuition of the essential qualities of the mesodermal elements in the nerve centers, because his description is applicable almost entirely to our microglia, albeit the cell forms analyzed by Robertson do not belong to the latter.

Microglia, an element of genuine mesodermal lineage, have all the attributes that move them away from the neuroglia, as Robertson's concept of this; that is, they are not part of the overall structure, are not related to blood vessels, nerve cells or fibers, and have phagocytic function. Instead, the oligodendroglia—to which belong the cells seen by Robertson, together with other new morphological types—constitute the main part of the neuroglial framework in the white matter of the brain and form special adventitias around the nerve tubes.

In summary, Robertson saw the small units of oligodendroglia with their characters barely sketched by the platinum method, and therefore he could not discover their fundamental morphological qualities and make a theoretical interpretation. Microglia were unknown to him (although he glimpsed some of their forms) but he, like many other researchers, suspected their occurence.

[1] Robertson's figure 26 in his text suggests that he saw a microglia element granularly stained.

## Cajal's Adendritic Corpuscles

Cajal (1913) happened to discern specific characteristics of oligodendroglia, although convictions about the universal affinity of the neuroglial expansions for the gold sublimate made him see the existence of indifferent round or apolar corpuscle as real and indisputable facts.

Since the oligodendroglia are absolutely refractory to the auric method, their elements show, in fact, colorless protoplasm and granulose nucleus. The formol-uranium method, on the contrary, provided images to Cajal in which the nucleus appeared clear and the protoplasm very dark. Precisely, these features of the apolar corpuscles are which approach them physiognomically to the oligodendroglia. Because of that, Cajal could see polyhedral, tuberous and mammillated elements and, from time to time, "a relatively long protrusion terminated by rounded clump." However, our master finds no evidence of the possible transformation of these elements in neuroglia. For our part, observing the same events with different criteria, we noted that formol-uranium dyes the expansions of oligodendroglia, giving the elements similar characters, except for the extent of protoplasmic appendages, to those shown after the staining with Robertson's methods, the silver carbonate and modified Golgi.

Everything about the distribution of oligodendroglia, identified as the "third element[2] of the nerve centers" or "Cajal's small adendritic cells" benefits significantly from the excellent description of our master.

Cajal states about neuronal satellites that are slightly rounded or tuberous, are 5–8 microns and gather in pleiad like eggs in a nest, living all around the neuronal contour and preferring the recesses of the base, where there are usually one to four and rarely more.

According to Cajal, the vascular satellites "are very numerous, offer variations in size and shape and often are arranged in series or parallel columns to the vascular axis. Are particularly concentrated in the angles of bifurcation, preferring the vicinity of capillaries to arteries and veins of any caliber. In short, more abundant in the vessels of the white matter than in the gray one."

The main site of the apolar corpuscles, according to Cajal, is the white matter, where they are either isolated or accumulated in series or groups of three or more elements and oriented along the dominant direction of the nerve tubes. In the formol-uranium preparations, Cajal saw the following types: (a) corpuscles 5–7 microns, with spherical nucleus and scarce protoplasm somewhat ovoid or polyhedral; (b) cells from 8 to 14 microns, with more abundant protoplasm mainly in the poles; (c) cells with protoplasm that begins two or three short or tuberous expansions; and (d) larger and richer in protoplasm corpuscles, with marginal excrescences

[2]Let us say again that the real third element of the nerve centers are the microglia, which have a mesodermal origin and specific qualities of morphology and function completely separate from those characteristics of nerve cells and neuroglia. Cajal's so-called third element is absolutely identifiable with the oligodendroglia; these, in turn, are a variant of the second element of the nerve centers. (See del Río-Hortega, Lo que debe entenderse por tercer elemento de los centros nerviosos. *Bol. de la Soc. Esp. de Biol.* 1923).

reminiscent of sprocket wheels. This type impressed Cajal about the possibility that such cells eventually may transform into astrocytes, but he rejected this gliogenetic hypothesis based on irrefutable arguments. Only one argument of those he poses fails, which is essential in our opinion: the alleged absolute specificity of the auric method for neuroglia. Moreover, the lack of morphological transitions with astrocytes, the nuclear characters, the low influence of age and gliosic processes in the number of elements, reported by Cajal, are indisputable facts.

Where Cajal's opinions were more accurate and shrewd is in regard to the relationship between adendritic corpuscles and the wealth of medullated nerve tubes and the resulting hypothesis that the adendritic cells of the white matter would be "something like a rudimentary Schwann corpuscle spread unevenly among the nerve tubes." Our master did not decide, however, to accept the homology of both types of cells, which could only be based on common pathophysiological features. As for the neuronal satellites, might be homologous to the subcapsular elements of ganglia and "could have been differentiated to fulfill near the neurons a special trophic role."

As noted already, despite the unfavorable observations that made him believe in apolar elements and assume their mesodermal origin, Cajal had a precious glimpse into the role entrusted to oligodendroglia.

So far we have made the critical summary of two publications that are epoch-making in relation to oligodendroglia. Our studies were done while we were ignorant of the first and bound to the suggestions and prejudices of the second.

Let our mentor, Cajal, see in our research on oligodendroglia the desire to investigate scientific truth without diminishing with blind proselytizing the prestige of his school and to collaborate modestly in a problem of his interest. And let our revered master see in this publication, that confirms one of his most astute insights, a tribute of loving respect.

## Initial Studies About Oligodendroglia

Our first suspicions that all elements of the nervous system called "apolar" by the authors had more or fewer branches began with the finding of microglia, which for the moment we thought identifiable with Cajal's "third element." Shortly after we were convinced that, in addition to microglia, other kind of corpuscles were much more abundant everywhere and showed marked preferences for white matter. We provisionally called it *glia interfascicular*.[3]

Understanding the importance of our study and despite technical difficulties, we decided to continue using the most favorable animals (dog, cat, and monkey) for staining small neuroglia corpuscles. Two years of research finally gave us the morphological clue of oligodendroglia, whose description we made in 1921, expanding rather setting the physiological hypothesis in 1922.

[3] P. del Río-Hortega, El tercer elemento de los centros nerviosos. *Bol. de la Soc. Esp. de Biol.*, 1919).

From the fragmentary observations seen in images, whether neat or confusing, we got a quite sure concept of the *oligodendroglia* or *glia with very few processes*, whose characteristics were shown to us as follows:

**a.** enormous number;
**b.** diffuse distribution in all regions of the nervous system;
**c.** great predominance in the white matter;
**d.** commonly grouped next to neurons (neuronal pleiads or satellites), following the course of the vessels (vascular satellites) and accompanying, arranged in long series, nerve bundles and fibers (interfascicular glia);
**e.** corpuscles small in size (often larger in the white matter) with round, angular, pyriform, fusiform, or stellate soma;
**f.** protoplasm dark and dense (sometimes light and loose) with few, very delicate, smooth or nodose, filiform or lamellar, long and slightly branched expansions;
**g.** nuclei round, thick, and clear;
**h.** cell body and its appendages with specific granules (gliosomes); and
**i.** the processes have a clear tendency to form faint discontinuous sheaths around nerve fibers, mimicking the reticular arrangements of Schwann cells.

Being well known the nucleus and the refractory-to-the-usual-dyes cuboid protoplasm, as well as the topographical disposition of this extensive system of gliocytes, the interest of our contributions lay in the demonstration of more varied types from those already known (especially in the interfascicular corpuscles) and in the findings of larger cytoplasmic processes than those described by Robertson, which were closely related to nerve fibers and formed complicated sheaths around the medullated fibers of the white matter of the nerve centers.

With respect to the cytoplasmic structures of such gliocytes, we demonstrated the existence of secretory granules in them that were identical to those present in protoplasmic and fibrous astrocytes. However, we erred in the description of cells with secretory activities by assuming there could surely be included in the oligodendroglia a special type of cell characterized by voluminous gliosomes existing in the marginal regions of the medulla and pons. Such cells with granules of extraordinary volume, visible especially in young animals affecting the period of intense myelination, seem to belong to a kind of glia whose morphological similarity to astrocytes or to oligodendroglia we have not been able to establish with firm basis so far.[4]

The irregularity of the results obtained in our study was the subject of great hesitation in the morphological description of the oligodendroglia, forcing us to make

[4]The brief summary of our papers from 1921 to 1922 did not mention much morphological or structural detail seen in the oligodendroglia, but these are thoroughly described in these publications. The 1921 publication's 12 plates with 19 figures demonstrated clearly the extent of our observations and were the best way to show our morphological concept of oligodendroglia to those who did not speak Spanish. Prints 1, 2, 3, and 4, however, aim to remember the adendritic aspects that oligodendroglia present with the methods of Cajal (gold sublimate and formol-uranium) and Nissl. With the best intentions, no doubt, our thoughts are distorted when either of these figures are copied to other publications, since they are aimed at demonstrating *how* the oligodendroglia *are not*. We have never believed that the oligodendroglia are as shown in Figures 4.3 and 4.8 of our work, although making us an honor we thank, they have been reproduced in the important books of Bailey and Cushing (*Tumours of the Glioma Group,* 1926) and Jakob (*Anatomie und Histologie des Grosshirns*, 1927)

errors of interpretation as previously noted. Instead, better research (based on stainings that could be considered complete if we were not already persuaded of the impossibility of staining the extent and limits of many protoplasmic processes) have given us neat images of structures seen earlier confusingly, have confirmed the overview of oligodendroglia and, finally, have made clear concerning their protoplasmic expansions and their relations with nerve fibers.

Thus, when comparing the images in this paper with those appearing in the former, a considerable difference in the overall morphology and interlocking of oligodendroglial elements is noted; these might even be considered different in some cases.

Our previous designs strictly conformed to the reality of 1921 without adding or removing protoplasmic branches from those that appeared in our preparations; instead, our description, based not only on neat and clear images but also on the blurry ones that we could discern, shows, after discussion and criticism of opposing images, that we were not satisfied with that seen until then in imperfect preparations achieved with fortunate techniques.

That is why during Dr. Penfield's attendance at our laboratory we made him interested in further research on oligodendroglia. We wanted a serious neurologist, hardly suggestible and a good technician such as Dr. Penfield, to check our findings and give his valuable opinion on obscure points.

## Confirmation and Discussion

Penfield's research (1925) fully confirms our original description and elevates interest in the topic of oligodendroglia in relation to classical neuroglia. Following the silver carbonate technique, with minor variations in detail, Penfield was able to verify the accuracy of our assertion about the existence of cytoplasmic radiations in all of the interstitial elements of the nerve centers, usually considered without them.

After analyzing the nuclear characteristics of the common neuroglia, microglia, and oligodendroglia, Penfield studied the protoplasmic expansions of the last, which are long, go across or run parallel to the nerve fibers, and appear flattened in the best preparations, forming incomplete covers for the myelin sheaths. Thus, the processes of the oligodendroglia originate irregular networks on myelin tubes, similar to Schwann cells. The American neurologist agreed with our hypothesis of possible involvement of the oligodendroglia in the formation and support of the myelin sheath. The hypothesis is based on the interfascicular situation of the cells and in the presence of fat gliosomes in the somatic and expansional protoplasm and during the period of maximal myelination.

Penfield finds remarkable similarity between the classical neuroglia and oligodendroglia, which, although of the same origin, are endowed with different functions and may not be seen transforming into each other in pathological cases. The only difference separating the two neuroglial varieties is the lack of vascular insertions in the oligodendroglia.

Our own and Penfield's research have been continued by many authors who analyze the properties of oligodendroglia from the morphological, topographical, physiological, histogenetic, and pathological points of view. However, the description of microglia and oligodendroglia is often made joint, although the former is more extensively treated (Metz and Spatz, 1924; Poldermann, 1926; Winkler-Junius, 1926).

Regarding morphology, Bailey and Hiller (1924) checked the general characters, in line with observations made with the platinum chloride, silver carbonate, and gold sublimate techniques. Creutzfeldt and Metz (1924) pointed out the similarity between lymphocytes and oligodendroglia and the close relationship between the latter and the microglia. Da Fano (1926) also verifies the morphological characters.

López Enríquez (1926) reported the characters showed by oligodendroglia in the visual pathways. Constituting an extension of the cerebrum, these should possess abundant oligodendroglia; in fact, López Enríquez has given clear proof by showing that it is arranged in interfascicular series up to 15 and 20 elements or spread unevenly in the optic nerve, chiasm, and tracts. Regarding the morphological characteristics, López Enríquez's observations agree with ours and Penfield's. The expansions are filiform or slightly flared, follow a more or less flexuous course, worming their way among the medullated fibers, emitting some collateral branches, and are lost without their ends being perceived in most occasions. Sometimes, however, after a more or less long tract and dividing like a T, end up widening and adapting along nerve fiber surface.

Also in 1926, but some months after López Enríquez, Marchesani described the oligodendroglia of the primary optic centers, optic pathways, and retina. The oligodendroglia appear in the optic nerve in close relation to the course of the nerve fibers and in the retina in relation to the thick ganglion cells. He believes that the predominant supporting elements are astrocytes.

In the morphological description of the oligodendroglia, Marchesani agrees with us regarding nuclear and protoplasmic characters. As for the processes, they follow different directions, the longitudinal ones seeming wider and longer than the transverse. López Enríquez thinks, according to our ideas, also accepted by Penfield and others that the processes form more or less complete covers to myelin sheaths and never settle into vessels, differing fundamentally in this from astrocytes.

## Contradictory Ideas

In general, given the insufficient techniques now used to study oligodendroglia, authors do not venture into full morphological descriptions, but some suppose to have found the secret of their processes. Among them, Urechia and Elekes (1926), studying the microglial syncytium, describe oligodendroglia and microglia junctions Bergmann (1927) returns to the neuroglial syncytium, observing transitional forms between neuroglial cells and granular bodies and noting the difficulty to establish the fuzzy border between neuroglia and microglia, which would be connected by intermediate forms; and Pruijs (1927) analyzes the boundaries between microglia and oligodendroglia and points out the great differences in shape and position of the nucleus, protoplasm, and ramifications. However, he describes intermediate borderline forms: cells which by the protoplasm and course of expansions correspond to oligodendroglia and, by the shape and position of the nucleus, to microglia.

In his excellent book *Anatomie und Histologie des Grosshirns*, Jakob gives deserved importance to the oligodendroglia and microglia, studying (with the efficient help of Uruguayan Schroeder) their normal and pathological characters. Although disagreeing with us on some points, it must be recognized that Jakob has managed to get from our contributions on microglia and oligodendroglia the maximum application to histopathology. Regarding the oligodendroglia, he describes their normal characters as they appear with the silver carbonate and analyzes their variations in pathological processes. Being a supporter of neuroglial syncytium, he includes the oligodendroglia in it, which, although they reveal scarce protoplasm, adapt to the glial reticulum.

Reynolds and Slater (1928) have done a good study of the neuroglia, and indicate, in accordance with our observations that astrocytes form fibers and have vascular feet, but not the oligodendroglia. At this point, however, there are discrepancies with Jakob's opinion, who considers oligodendroglia able to undergo a fibrous transformation, and with that of Bailey and Schaltenbrand, who admit the existence of vascular feet in oligodendroglia and describe their participation in the *membranae limitantes*. These authors believe that these cells are transforming into astrocytes, although these, except for the vascular foot, show all the peculiarities of the oligodendroglia, in which the development of widened vascular pedicles is not as complete as in the neuroglia.

Regarding the histogenesis of microglia, although their identification as Robertson's supposedly mesoglial elements and as Cajal's third element, also suspected of mesodermal roots, could have influenced the authors' mind, when dealing with histogenesis of the interstitial elements of the nerve centers, they set the oligodendroglia in the appropriate place—that is, among neuroepithelial ectodermal elements. This is seen in the classifications of Penfield, Bailey, Cushing, and so on. Penfield subscribes the ectodermal origin of the oligodendroglia, considering them a specialized type of neuroglia, not germinal or undifferentiated. Reynolds and Slater (1928) have also studied with regard to histogenesis, admitting that the oligodendroglia come from glioblasts without processes and with ectodermal origin, whereas microglia probably originate from mesodermal tissue of the pia mater. Reynolds and Slater find the oligodendroglia especially rich in young animals at the time of myelination in which the cells appear very granulose.

According to Bailey (1927), the oligodendroglia derive from pluripotent medulloblasts, which can also originate unipolar spongioblasts or unipolar neuroblasts, the latter giving rise to nerve cells.

Bailey and Hiller (1924), with conflicting ideas from Robertson, Cajal, and us, did not decide to include the oligodendroglia within neuroglia, grouping microglia under the question "mesoglia?" Bailey's ideas are now more accurate and close to ours, since in his most recent collaborations with Cushing and Schaltenbrand includes the oligodendroglia among ectodermal gliocytes.

The authors cited to this point agree with all or part of our ideas on oligodendroglia, but some researchers do not.

Among the authors who hold opinions contrary to ours, we must mention Cajal, who in 1920 insisted on the existence *of dwarf* or *globular satellite corpuscle* resistant to staining and apparently devoid of appendages, far from both the satellite microglia as the perineuronal macroglia. In his *Manual de Histología* Cajal (1928) maintains the same ideas.

Schaffer admits in 1926, according to Cajal, adendritic astrocytes, part protoplasmic and part fibrous in nature and adendritic or a differentiated third element.

## On the Functional Concept

Studies by Schaffer in pathological material leads him to hold in harmony with Pollak that the apolar neuroglial or third element normally remains dormant but ready to intervene (*Bereitschaftenszellen* of Pollak) in exogenous and endogenous parenchymal diseases. Then the "prepared cells" become active, proliferate around the pathological material, incorporate fatty products, and, like true emigrant glia (*Wanderglia*), reach the perivascular lymphatic clefts and deposit the fatty material in them. This makes clear that in this conjecture Schaffer involves and confuses two divergent concepts: that of oligodendroglia (as seen by him without processes) and that of microglia.

It is not the Hungarian professor, whose opinions deserve all our respect, the only one who attributes to the oligodendroglia activities that do not perform in any way. His disciple Meduna (1927) tried to do as well (with little success in our opinion) a division of roles between microglia and oligodendroglia, interchanging them in a way that conflicts with everything shown by numerous authors. According to Meduna, microglia (a variety of glia) would be capable of swelling in pathological cases, but the oligodendroglia would, in addition to other activities, have the ability to phagocytose the degenerated microglia (*microgliophagy*).

The possibility of this phenomenon is comparable to that of neuronophagy. When ignoring the existence of microglia and without showing their peculiar phagocytic activities, it was natural to speak of neuronophagy performed by hyperplastic neuronal satellite elements, although in small round cells no evidence (protoplasmic inclusions) of such phagocytosis was found; but returning now on that topic without providing any evidence or clues, and to attribute to the oligodendroglia phagocytic roles that will never be proved, seems a mistake to us. We have said emphatically that the only specific phagocytic macrophages of the nervous centers are the microglia and we cannot accept any report opposing this firmly grounded idea, if not accompanied by irrefutable proofs.

Although some researchers on microglia were not decided to accept our assertions and admitted (by the phenomenon of inertia) the possibility that some neuroglial elements would originate granulo-adipose bodies, their most recent publications attest to the evolution of their ideas, each day closer to ours.

In their research on multiple sclerosis, where oligodendroglia contain small clusters of fatty droplets, Creutzfeldt and Metz (1926) never saw the formation of *Gitterzellen* nor *Kornchenzellen* from oligodendroglia. The oligodendroglia would weakly participate in the retention of metabolic products.

Struwe (1926) has done a beautiful study about the fat content of the three neuroglial types, noting that the oligodendroglia and astrocytes may have lipoid accumulations in their cell body, but only microglia are able to form free granulo-adipose bodies. According to Struwe, the oligodendroglia participate moderately, mostly in the *Fettspeicherung*, showing clusters of granules on one side of the nucleus, on opposite sides or around (Creutzfeldt). If the amount of granules is large, they form striae radiating from the nucleus. The combination of silver carbonate techniques and Scharlach shows that the fat lies in the fine processes and in their vesicular thickenings. Struwe believes it possible that astrocytes and oligodendroglia are active in nervous demolition in a different manner from the microglia but makes no final judgment.

Previously, Metz (1926) demonstrated that the morphological differences of astrocytes, oligodendroglia, and microglia correspond to a different behavior in the storage and transportation of materials, an idea which in principle is not opposed to ours that only microglia are able to retain and transport exogenous materials, in disagreement with that of Pruijs (1927), who considers it possible to form granulo-adipose bodies, at the expense of both microglia and oligodendroglia. Also, Jakob (1927) expresses a similar opinion on accepting that all glial forms may be involved in destruction and become granular bodies, placing first microglia and oligodendroglia. Opposing this idea are the observations of Collado, Spatz, Metz, Creutzfeldt, Alberca, Penfield, Asúa, and so on for whom the microglia are the only elements phagocytically active and therefore able to engender granular bodies.

The most important studies related to the physiological activities of the oligodendroglia were carried out by Metz and Spatz, Struwe, and Malamud and Wilson and refer to the binding properties and retention of iron (*Eisenspeicherung*) in normal and pathological conditions.

Continuing Spatz's all-important research on cerebral iron, Spatz and Metz undertook the study of their location in the two cell types described by us. In 1924, they published their observations on microglia, and in 1926 those on oligodendroglia. Metz's research opens the way, revealing that the oligodendroglia have the property of fixing iron in the vicinity of the foci of softening and hemorrhage—that is, when there is a simple disintegration of nerve tissue and when ferric foreign bodies are present. Later, Metz and Spatz studying the general paralysis, noted that the normal property of storing iron by oligodendroglia is dramatically patent in paralytic dementia, a finding of histological importance.

As described by Metz, oligodendroglia contain very fine ferric granules, the cell body often showing a diffuse bluish staining. Pathological conditions can increase the amount of blue granules, which are fine, round, about the same thickness and placed in groups or scattered around the nucleus or away from it. Occasionally, from the periphery of some cell series, granulose lines irradiate, marking the course of the delicate expansions. The fixing of iron studied by Spatz in different brain areas, according to Spatz and Metz, would take place in the oligodendroglia essentially, microglia participating in the *Eisenspeicherung* when the abnormal iron content reaches a high level. When cerebral iron (*Gehirneisen*) increases, appears almost entirely in the ectodermal parts of the tissue; on the contrary, the paralysis iron (*Paralyseeisen*) is located in the mesodermal elements of the vessels and in microglia, but not in the ectodermal varieties of glia; in the eventual existence of blood iron (*Blutzerfallseisen*), astrocytes and oligodendroglia are involved in fixing it, but microglia have the main role.

Struwe's research, which compared the situation of fat and iron in the oligodendroglia as an extension of Metz's study, show that the iron granules are usually thinner and more widespread in the cell body than lipoid granules. Malamud and Wilson recently continued the research on iron, making observations similar to those reported on the softening foci in the ferric substance deposits and in general paralysis. The oligodendroglia, as Struwe noted, contain fine iron granules but to a lesser amount than microglia.

According to the publications cited, the oligodendroglia can accommodate lipoid and ferric granules in their protoplasm, corresponding to their capability to intervene in degenerative processes and physiological or pathological iron storage.

While these activities based on the results obtained with specific methods for fat and iron do not seem controversial, there is in them a major dilemma: Is the ferric or lipoid content exogenous or endogenous? Does it correspond to a protoplasmic special processing or to extracellular product adsorption or absorption phenomena? Winkler-Junius's (1926) observations, by failing to demonstrate traces of active modifications in macroglia, make him to consider very possible that eventually there were in them lipoid granules passively uptaken or as a result of cell degeneration. On the contrary, microglia would behave as a phagocytic mesodermal element.

Without knowing well the morphology of neuroglia with few processes and without knowing exactly the relationships they have with nerve cells and fibers, especially the latter, their function cannot be understood. This is not phagocytic, of course, because macrophage functions are the sole province of microglia. Apart from the possibility that oligodendroglia retained products of their own metabolism, oligodendroglia could also elaborate special substances, as evidenced by the existence of secretory granules in their protoplasm.

## Relations Between Oligodendroglia and Nerve Fibers

Regarding the homology of oligodendroglia and Schwann cells, defended by us and in substantial agreement with the ideas of our master, there are some observations in the literature of great interest.

On dates before and after the already mentioned publications, many others have dealt directly or indirectly with oligodendroglia, providing interesting data regarding location, shape, microchemical reactions, and pathological changes. Some of these publications relate to the common neuroglia in normal or pathological state, but in the figures linked to their description is clearly seen that also correspond in some part to oligodendroglia, from which interesting protoplasmic formations were seen and sometimes analyzed in detail.

The oldest of these publications and the most important is that of Paladino, who reveals the connections between neuroglia and nerve fibers. In effect, Paladino in 1892 made the interesting observation that neuroglial cells of very different dimensions sent processes among the fibers of spinal cord funiculi and even within themselves, forming through the myelinic area a reticulum that came to encircle the axis cylinder.

With the participation of fibers near to or far from a gliocyte body, a true neuroglial skeleton would be engendered in the nerve tubes. Later (1911), Paladino extended his original concept, arguing that the neuroglia continue in the myelin sheaths of white and gray matter, forming a more or less complex skeleton with spiral fibrils coursing in varied directions.

These interesting observations were echoed by several authors, including Nemiloff (1910), who confirms the neuroglial nature of Paladino's network. Marano accepts the penetration of the neuroglia in the medullary sheath. Perusini (1912) discloses the variegated aspect of the neuroglial skeleton in the different areas of the same spinal cord section. Doinikow (1911) and Jakob (1912) comparatively analyze the alterations of Paladino's reticulum in the spinal cord and skeleton of Schwann cells in nerves, noting absolute parallelism. Finally, Montesano (1912) brings new and important observations on the constitution of Paladino's *schcletro nevroglico*.

Since the formation of peri- and endomyelinic structures of nerve tubes was attributed to the classical neuroglia, it could hardly be known to what extent the researchers observed the involvement of the oligodendroglia in that process; besides, the images copied by them show misleading features. But once we saw that the oligodendroglia are precisely the unique variety of neuroglia surrounding and walling up the myelinic sheaths, we believed that Paladino together with those who subscribed his concepts could discover, although with imprecise characteristics, part of the formations belonging to oligodendroglia. And it offers no doubt when you see that certain subependymal elements with processes related to epithelial cells copied by Paladino have similar features to those of oligodendroglia; that the interfascicular gliocytes of the spinal cord, forming long series in relation to nerve fibers but not with vessels, as described by Perusini, are obviously glia with very few processes; and finally, some cells copied by Montesano (Figures 1, 7, and 10 of his publication) belong certainly to medium-sized oligodendrocytes with processes around nerve fibers. However, most of the observations previously described correspond to fibrous astrocytes whose expansions are added to the general neuroglial plexus but not more associated to the true peri- and endomyelinic formations.

Because of this, when Montesano found in the spinal cord's white matter elegant perimyelinic formations, numerous fibrillar structures of various sizes (some of which are related to neuroglial cells), and filamentous apparatuses in the most diverse direction in spiral, networked, like volutes, which come from adjacent or remote cells, we think they refer exclusively to oligodendroglia and do not hesitate to admit that Paladino and Montesano saw more important details and made more accurate interpretation than Robertson with this variety of neuroglial cells. Here is why the current concept of the oligodendroglia is opposed at all to that Robertson had of his mesoglia, but agrees substantially with that of Paladino about the neuroglial skeleton. Robertson denied the existence of any relationship between mesoglia and nerve fibers. Paladino, in turn, showed intimate connections between neuroglia and myelinic tubes. For our part, we intend to prove in this work that Robertson's mesoglia form delicate sheaths on all nerve fibers and that Paladino and Montesano's peritubular filamentous apparatuses are also oligodendroglia formations.

Paladino was the first to observe the peritubular neuroglial formations, but we note that their conclusions are based on very fuzzy images, in which, making a critical analysis, could not be clarified the share of simple neuroglia and oligodendroglia. However, it can be reasoned that most part of the intertubular neuroglial formations described by Paladino and Montesano belong to the fibrous neuroglia but do not correspond to the intimate sheaths of nerve tubes. The astrocyte processes simply follow longitudinal or transverse paths that are more or less wavy.

In the studies under discussion, there are not only hits but also errors of some importance, such as the very significant regarding the presence of neuroglial cells in the space between the perimyelinic fibrillar structures and the axis cylinder. In this regard, Paladino says that between the axis cylinder and the periphery there are small nuclei to which converge branches in several directions. Montesano, however, says that the endomyelinic cells are different from those forming neuroglial sheaths: They are significantly smaller and have scarce, round, spongy, and process-free protoplasm; that is, with characters similar to Cerletti's *piccole cellule* and Jakob's *protoplasmaärmste Zellen*, now identifiable with the oligodendroglia, which tacitly denies, therefore, any involvement of oligodendroglia in the genesis of neuroglial spirals.

These results demonstrate the limited accuracy of the stainings obtained by the referenced authors, who attributed to the neuroglial astrocytes the filamentous apparatuses originated by the oligodendroglia, which were located partly within the myelin.

## Abnormal Variations

Regarding the oligodendroglia abnormal states caused by cadaveric events or true pathological disturbances, there are already quite interesting publications. The oldest are from Grynfeltt and Pélissier and are referred to as the so-called mucocytes of the nervous system.

Grynfeltt pointed out in 1923 the presence of special lesions in the cerebrum of an old man with circumscribed edema foci. They consisted of light spots of neat contours, rounded or polycyclic, and full of metachromatic substance with the mucin reactions. This mucin would result from the degeneration of certain cells (mucocytes) scattered in the white matter, among myelin fibers, being particularly numerous in the perivascular foci of edema. Mucous substance that also exists in the subependymal region among epithelial cells of the ependyma and even in the ventricular cavity, Grynfeltt supposes to be evacuated in the subarachnoid spaces and ventricles.

Continuing his mentor's studies, Pélissier (1924) stresses the issue of mucocytes, identifying them with the interfascicular glia described by us. In the circumscribed edema foci in a case of Wilson's syndrome, he finds reasons to entirely confirm Grynfeltt's finding and adds new details on mucocytic degeneration. In the serial elements of the interfascicular glia would appear dense droplets of a mucosal substance that would coalesce and dilate cell contours, then mucocytes themselves merge, forming masses of mucus in which degenerated nuclei would float; finally, the evacuation of mucus, as imagined by Grynfeltt, would take place.

This interesting degeneration of the interfascicular and perivascular oligodendroglia, autochthonous "necrobiotic fusion",—that is, independent, as Pélissier, of nervous breakdown phenomena, to reach full confirmation, would have obvious importance. However, final acceptance would need a demonstration avoiding Bouin's fixative and desiccation, factors little favorable to neural structures.

Because we neither favor nor oppose observations to oligodendroglia mucocytic degeneration, we can only think in terms of the technique followed, which we consider insufficient for studying the process. Identical objections could be done to Bailey and Schaltenbrand's most recent contributions about the previously noted mucoid degeneration of the oligodendroglia.

Bailey and Schlaltenbrand (1927), studying a case of Schilder's diffuse periaxial encephalitis,[5] found foci of mucoid degeneration, largely confirming Grynfeltt's description.

Such degeneration was not visible in the regions most severely destroyed, only in those areas where the disease process began. Observing the origins of the process, Bailey and Schaltenbrand saw that the protoplasm of small, round glial elements, first granular and with the characteristic configuration of oligodendroglia, began to swell and become stainable with mucicarmine; afterward it lost the granules, swelled and got clear and took uniform red color; finally cell borders disappeared, leaving the mucin in the tissue environment.

The mucocytes appeared as clear spots scattered throughout the white matter, isolated or in clusters, containing a substance stainable with hematoxylin in reddish blue, with thionine and toluidine in metachromatic red and with mucicarmine in bright red—that is, with the characteristic reactions of undefinable mucous matter.

Baily and Schaltenbrand say they are not astrocytes or microglia, but true oligodendroglia, as seen by comparing the sections stained with mucicarmine, silver carbonate, toluidine, and phosphowolframic hematoxylin. In the silver impregnation, appear exactly the characters seen by Penfield and Cone in the acute swelling

[5]In a communication presented at the Academy of Medical Sciences of Barcelona on May 10, 1927, and published in July that year (*), speaking of the gliosis characteristic of Schilder's disease, we said the following (page 91): "In the interesting pathological process of Schilder's disease not only surprises the total disappearance of the medullated fibers in the affected parts, but also the almost complete lack of oligodendroglia, with a slight increase in the amount of interfascicular fibrous neuroglia, whose fibrillar appendages become numerous. Accepted by us the homology between neuroglia of few radiations and Schwann cells; assuming that the former plays a passive role of isolation and support, and other active of trophic order, which may be related to myelin genesis, properly studied, an argument for this assumption could be found in this parallel disappearance of myelin and oligodendroglia. For our part, we could only study a case of Schilder's disease, thanks to Dr. Stevenson's kindness, and we cannot leave the field of conjecture."

(*) "Gliosis subependimaria megalocítica en la senilidad simple y demencial". *Revista méd. de Barcelona*, 1927, año IV, t. VIII, núm.45, págs. 85–98.

of oligodendroglia, so it is identifiable with Grynfeltt's mucocytic degeneration. According to Bailey and Schaltenbrand, they are varying degrees of the same process.

Based on the study of many pathological cases and numerous experiences, in 1926 Penfield and Cone made an important contribution to knowledge about the abnormal changes in the oligodendroglia, describing the acute swelling as a specific type of injury as evidenced by silver carbonate. This lesion, which had been vaguely seen by Rosental, was characterized by hypertrophy, swelling, and vacuolization of oligodendroglia cells. The phases of the process are irregular increase of the perinuclear cytoplasm, with granular appearance of the expansions; hydropic swelling and progressive vacuolization of the soma, in which an unstainable fluid rejects the cell membrane, the surface and protoplasmic septa appearing granular; a decrease in the number of expansions, which are represented by strings of granules; and finally, in extreme cases, breakage or disappearance of the membrane and release of the nucleus, shrunk, and pyknotic.

Penfield and Cone find such lesions in various pathological processes of the nervous system and other locations (pneumococcal meningitis and tuberculosis, meningeal sarcoma, poliomyelitis, hydrocephalus, stroke, abscess, myxedema, heart failure, malnutrition, chloroleucosarcoma, and gastric cancer) but not in cases of traumatic or surgical death (brain tumors, stroke, and epilepsy). As for animal experimentation (emaciation, wounds, arterial ligation, softening, guanidine hydrochloride, ether, morphine and chloretone intoxications, intravascular injection of distilled water), causes generally more or less intense acute swelling of oligodendroglia.

The interpretation of the phenomenon studied by Penfield and Cone as real injury is difficult from the moment it is known to occur postmortem on autolytic basis, identical to the clasmatodendrosis, and the emaciation, the stupor and agony favor its existence. Moreover, the reiterated fact of microglia keeping intact in pathological processes where the acute swelling of oligodendroglia reaches its maximum, and microglia also remaining unchanged in many experiences, seem not to be linked only to the ectodermal or mesodermal lineage of the respective elements.

Knowing that the global or molecular disintegration of nerve tissue is not only manifested by neuroglial reactions but also by microglial alterations, the absence or presence of the latter may have great value in interpreting Penfield and Cone's acute swelling as a real lesion as engendered by a chemical–physical mechanism.

The described lesions of the oligodendroglia remind us of observations we published in 1909 when our ignorance of the neural histopathology made their accurate interpretation difficult. Apropos of a case of cerebral hydatic cyst, we described the considerable wealth of swollen neuroglia bodies existing in the white matter, which were stained in reddish purple with thionine and, because they were just being formed, appeared somewhat granular (myelinoid granules), but soon their protoplasm swelled, rejecting the nucleus and acquiring a rounded, ovoid, or scalloped shape and becoming homogeneous in appearance. In making the interpretation of these elements, whose resemblance to Grynfeltt's mucocytes is evident, we were right at considering swollen neuroglia, but we were wrong in calling them *myelophages*, assuming that stored granular or molten myelin. Today they could be included in abnormal forms of the oligodendroglia, as neither the common neuroglia nor microglia offer similar appearances. They are swollen corpuscles filled with a substance stainable in red with thionin, just like the mucocytes recently studied by Grynfeltt, Pélissier, Bailey, and Schaltenbrand.

Besides these variations, which corresponding either to real degenerative processes, either simply to deformation due to protoplasm tension changings, ameboid-type alterations and even secondary fibrillary degeneration (Jakob) have been identified in oligodendroglia. It is often also described the formation of granular bodies, which means that, when they both are swollen, a clear distinction between microglia and oligodendroglia has not been achieved yet.

The hypertrophy and hyperplasia of oligodendrocytes in encephalopathies, although described and demonstrable in many cases, are not precisely assessed. So Jakob (1927) did not decide to make a definitive judgment regarding the ability to proliferate, and Alberca (1928) pointed out the difficulty of recognizing the quantitative alterations. However, although the classic concept of neuronophagy (*normal pseudoneuronophagy* after Spielmeyer) is no longer applicable, the memory of the images in which it was founded remains.

Schaltenbrand and Bailey (1928) argue that, in conditions still undetermined, the oligodendroglia can proliferate diffusely. In Banti's disease, chronic malaria, and other diseases, these authors observed lots of small round neuroglial cells that tend to form perivascular walls and neuronal satellites (*satellitosis*).

In the study of brain gliomas, the knowledge of oligodendroglia, although still deficient, has pointed in new and brilliant directions to Bailey and Cushing (1926). Under the heading of oligodendrogliomas, these neurologists describe (with maximum accuracy that allows the imperfection of the techniques) the tumors formed by the proliferation of oligodendroglia and characterized by having two types of cells, which appear as a tangled web with the gold method: oligodendroglia corpuscles, round, with clear cytoplasm and nucleus rich in chromatin, and some scattered astrocytes.

The group of oligodendrogliomas does not yet seem definitively specified, judging by the latest contribution of Bailey and Schaltenbrand (1927), in which a question mark after the name is set as a sign of uncertainty.

The foregoing critical review of the literature dealing with oligodendroglia (by that or another name) is perhaps too long but we fear we have not referenced some interesting publications that are not in our modest laboratory library.

# 3 Material and Technique

Our research has taken place mainly in the nerve centers of man, monkey, dog, cat, and rabbit. The human material belonged to individuals who died as a result of various diseases, especially those of the nervous system. Mental disorders and encephalopathies studied are general paralysis senile dementia, Huntington's chorea epidemic encephalitis, Schilder's periaxial encephalitis, meningoencephalitis, syphilis, tuberculosis, tumors, hydrocephalus, and so on.

By giving a general description of oligodendroglia, we do not intend to make a complete analysis of their alterations regarding location, quantity, and quality, rather we want to present some observations related to the possibility that they may be involved in the pathological substratum of various diseases manifesting specific forms of reaction and degeneration.

Our object is now limited to describe oligodendroglia as they appear in normal state.

As the rabbit is the animal that best demonstrates the microglia, the study of oligodendroglia is more easily accomplished in the monkey, cat, dog, sheep, and donkey. Human material can give excellent results if it is picked up shortly after death occurs; between 16 and 24h after death can provide estimable preparations, both with silver carbonate and the technique we will describe later.

Undoubtedly, to achieve high demonstrative efficiency stains, the use of elective metal impregnations is required, which, although reproachable for their inconstancy, can never be surpassed in beauty by those obtained with anilines. Generally, we use silver methods and always recommend them, but not exclusively, and also, as a practice, we compare these images with those stained with Nissl method. This control has persuaded us of an important fact: toluidine blue and thionine, well managed, are excellent resources for oligodendroglia research in pathological cases. They resolve nothing with respect to normal or pathological morphology but give very sharp images regarding quantity, location, and progressive or regressive state of oligodendrocytes. Because authors are now studying the pathological states of oligodendroglia, it seems necessary to examine anew those older specimens prepared with the Nissl method; it will permit us to compare the differing aspects shown with this method with those obtained with our silver carbonate or a Golgi modification, which have been the basis to carry out the present work.

The toluidine blue not only reveals the nuclei of two neuroglial varieties (astrocytes and oligodendroglia) and microglia but also show, although faintly stained, the protoplasm of the abnormal microglia (which thus could be seen by many authors since Nissl discovered the *Stabchenzellen*), drawing, sometimes quite accurately, the outline of oligodendrocytes when they are abnormally swollen. The protoplasmic mass is hardly discernible in cases of absolute normality if the fixation took place in

**Río-Hortega's Third Contribution to the Morphological Knowledge and Functional Interpretation of the Oligodendroglia.**
**DOI: http://dx.doi.org/10.1016/B978-0-12-411617-7.00003-1**
© 2013 Elsevier Inc. All rights reserved.

good conditions; but on the contrary circumstances, stand out very clearly, the neat contour, and the typical rounded or polygonal shapes, and the clusters of perineuronal and perivascular oligodendroglia elements.

A study of this neuroglial variety, to be complete, should not ignore the nuclear scan, either with the Nissl method or its derivatives, or with metal impregnations. The nuclear variant of our technique to silver carbonate is quite useful, since it completes the results of staining the protoplasm with the corresponding variant of the same method.

The silver carbonate technique, conducted according to known standards or following new and more effective adaptations can improve the results of previous research, but it still has well-known difficulties that must be overcome by testing the time of performance and concentration of the silver reagent. Following this practice, and by using the most adequate animal tissue, that of dog and cat brain, we obtain a clear display of oligodendroglia, sometimes almost completely impregnated.

An effective use of our technique requires that the tissue be fresh and not damaged, since the sensitiveness of protoplasm is not immune to cadaveric autolysis and rough handlings.

Our first technique subjects, more or less strictly to the following standards:

1. Fixation in formol bromide, 12–48 h.
2. Heating the pieces in formol bromide, 10 min at 45–50°C.
3. Freeze sections.
4. Wash in water with 10–20 drops of ammonia and then distilled water.
5. Impregnate for 5–15 min in strong solution of silver carbonate:
   10% silver nitrate solution: 5 cc
   5% sodium carbonate solution: 20 cc
   Ammonia, enough to dissolve the precipitate, avoiding excess.
6. Reduce in 1% formaldehyde, or after a very quick wash in 10% formaldehyde.
7. Auric toning and fixation in sodium thiosulphate.

Any changes in this technique consisted in treating the sections with 95% alcohol (2–24 h) before the silver impregnation or after it (30 s), preceding the reduction in formaldehyde. The alcohol quite favored the protoplasmic stain.

While attending our laboratory, Dr. Penfield found the benefits of alcohol use as a fixation complement. After washing the pieces during the right amount of time to eliminate alcohol and then freeze sectioning, Penfield followed the silver carbonate technique, preferring our first formula, slow acting, to the more concentrated and faster acting.

Using the techniques already mentioned a fully satisfactory staining of oligodendroglia is not yet possible. We needed to test many variations until finding one likely to give excellent colorations. For this reason, given the well-known advantages of hypnotics for displaying nerve and neuroglial structures, we have used some of them together with formaldehyde. The veronal, the sulphonal, the dial, the luminal, etc., can be used, but we have a preference for chloral hydrate, used as follows:

10% formalin: 100 cc
Chloral hydrate: 1–2 g

To this formula, in which the quality and quantity of hypnotic can be varied, it is sometimes useful to add 0.50 g of urea or 1 g of ammonium bromide.

One helpful technique is:

1. Fixation in the previous liquor for 2–12 days.
2. Washing in ammonia water.
3. Impregnation in concentrated silver carbonate 10–15 min.
4. Washing in 60% alcohol 1–2 min.
5. Reduction in 10% formalin.

Using this method in dog and cat brain we could have attempted a description of the oligodendroglia that would be more accurate than the one in our earlier works and those of Penfield, López Enríquez, Marchesani, and so on, but there would still be dimly perceived fundamental structures that are shown with absolute clarity with another procedure that we can now consider. This technique is a modification of Golgi's, by means of which the images of oligodendroglia appear as clean, neat and demonstrative as those of neuroglia cells of short and long extensions obtained with the silver chromate classic method.

This technique, which we may soon work to perfect, is the basis of this publication. We could have also used the technique to our study, platinum chloride of Robertson, if not offered as proven implementation difficulties and whether images of the oligodendroglia exceeded those obtained by the author. However, since it does not reveal anything but the cell bodies and beginning of some extensions, just divided, of the oligodendrocytes, it is easily surpassed in choice and constancy by this very simple technique:

1. Tissue fixation (2- to 3-mm-thick pieces) for 2 or 3 days, in the following liquid, which can be renewed daily:
   potassium dichromate, 3 g
   chloral hydrate, 2–5 g[1]
   10% formalin, 50 cc.
2. Quick wash in distilled water.
3. Impregnation in 1.5% silver nitrate for 2–3 days.[2]

[1] May be replaced by sulphonal, veronal, dial, and so on.

[2] *Important notes*: First, staining of the slices is very pale and yellowish; so, the occurrence of stained elements should be checked under a microscope before considering the technique failed. Second, the selective reaction is limited to a thin thickness; so, there are usually not more than 10–20 useful slices, whose useful areas are usually more reddish in color than others. Third, being very delicate cell processes and the observation needing to be made at high magnifications, the mounting must be made as soon as the presence of small, rounded or polyhedric, corpuscles is recognized in the slices, at low magnification. Fourth, to finish the preparations, we follow this routine:

a. Dehydration in 95% alcohol.
b. Clearing in creosote on the slide.
c. Drying with filter paper and mounting with balsam or damar as usual in the Golgi method.
d. Desiccation at gentle heat until complete evaporation of xylene.
e. Coverslip at room temperature.
f. Heat again until damar resin fusion.
g. Press with finger or a blunt object to remove any excess balsam, considering that the amount of balsam should not be higher than in a normal preparation, to use the immersion objective.

The material must be very fresh and sectioned, without bruises, into 2-mm-thick pieces.

The above written technique provided us the best stains among the many fixative formulas we have tried, varying the proportion of hypnotic (1–10%) and dichromate (1–5%) and using as vehicle water, 10% formalin or Cajal's formol bromide. The addition of pyridine (10–12 drops) favors only the staining of nerve cells.

The dichromatation should not continue to hardening of the pieces, since the effect is achieved when the tissue is thoroughly impregnated (1–3 days) and does not improve afterward. Regarding the action of silver, nothing is gained after 3 days.

The practice of renewing once the fixative and silver liquids is desirable but not essential.

The stainings are not specific, since besides oligodendroglia, protoplasmic and fibrous astrocytes often appear mainly perivascular. It is also common to see very finely impregnated nervous elements with broad ramifications. Purkinje cell branches and cerebellum grains appear very often. Axis cylinders never stain.

The oligodendroglia occur impregnated in black or brown color, not the full extent of the slices, but piecemeal, as is the rule in the reactions of the Golgi method.

In the white matter, it appears far better than in the gray, because it requires that the fixative liquid to penetrate in the expansional protoplasm very subtly. Thus, it is still an imperfect technique but always stains oligodendroglia cells in a higher or lower number, sometimes including all their processes and sometimes only some of them.

The best results are obtained in the cat and dog, young or old. The rabbit usually gives incomplete pictures. We have not been able to use monkey material, which for neuroglia studies is extremely favorable. In human material collected from 16 to 24 h after death, it has provided us with oligodendroglia staining that, although rather poor, is sufficiently demonstrative to enable us to compare them with findings made in laboratory animals.

# 4 Morphological Characteristics of the Oligodendroglia

If the name *oligodendroglia* corresponded to a cell type as morphologically uniform as protoplasmic or fibrous astrocytes, then their overall description could be done without difficulties, but the reverse is happening—and being very large, the polymorphism of oligodendrocytes (Figure 4.1) almost must be forced to describe their main aspects separately. There are, however, common features in all of them, about which an examination of a whole may be done.

The elements are highly variable in size, round, polyhedral, and pyriform in shape, having a round nucleus surrounded by scarce protoplasm. From it originates a variable number of long and soberly divided processes. Defined in this way, they could not be distinguished exactly from the protoplasmic and fibrous astrocytes, with which they show both analogies and important differences.

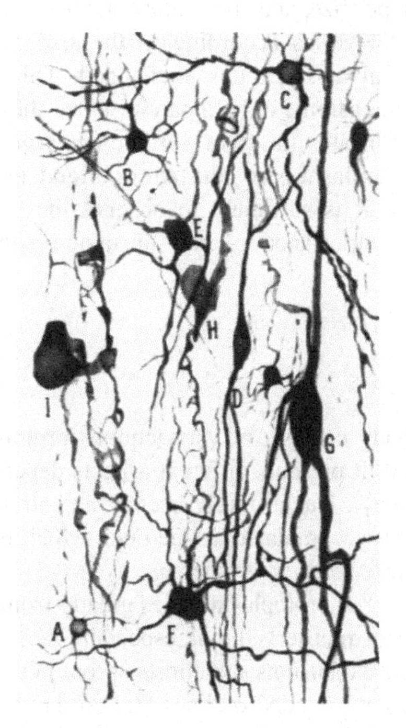


**Figure 4.1** Varieties of oligodendrocytes in the white matter of the dog brain: A and B, first type (or of Robertson); C–F, second type (or of Cajal); G–I, third type (or of Paladino).

**Río-Hortega's Third Contribution to the Morphological Knowledge and Functional Interpretation of the Oligodendroglia.**
**DOI: http://dx.doi.org/10.1016/B978-0-12-411617-7.00004-3**
© 2013 Elsevier Inc. All rights reserved.

## Nuclei

Comparing the oligodendroglia and astroglia nuclei in toluidine blue stainings, the size is of little value, since there are large oligodendrocytes whose nuclei are equal in size to those of astrocytes. One can see that the nucleus of the latter is slightly ovoid containing a tiny amount of chromatin in a clear karyoplasm, whereas the nucleus of the former is round and rich in chromatic lumps, similar in appearance to lymphocytes and transitional forms of these to plasma cells and (among corpuscles of identical embryonic origin) to the cerebellar grains. Therefore, it is not surprising, it has often been believed lymphocytes corresponded to the bare nuclei, and with conviction due to the fact they exist between the series and perineuronal groups of oligodendrocytes microglial elements, whose nuclei are even darker due to greater abundance of chromatin, only differing from lymphocytes in shape, when elongated, but not when round.

Gliocyte nuclei show size and chromatin rich gradations, thus very different at the ends of the series, become confused in the middle, where the distinction between astrocytes and oligodendrocytes becomes difficult. The eye accustomed to observing pure nuclear stainings easily recognizes the corresponding neuroglia types, unlike in the perception of the oligodendroglia elements among the cerebellar grains whose nuclei show identical shape, size, and chromatic content.

The size of the nuclei varies according to the size of the elements, yet the nucleocytoplasmic ratio always remains unchanged. This is characterized by an imbalance of nuclear dimensions, only apparent, since the development of expansional protoplasm compensates the poor perinuclear protoplasm. The mass ratio between nucleus and protoplasm is so constant in oligodendroglia (and in the other neuroglial varieties) that it is sufficient to observe the size of their bare nuclei and thus imagine the importance of the unstained somatic and expansional protoplasm.

## Soma

The oligodendrocytes body consists of very tenuous protoplasm virtually resistant to all dyes. When stained, it presents different aspects depending on technique used and better or worse success. Nuclei rarely occupy a central position, whereas protoplasm generally appears accumulated at one pole, which in many elements, is the place from where the thicker processes emerge.

The polarizing tendency of protoplasm varies greatly from one element to another and causes the polymorphism observed in it, especially those lying in the white areas of the brain. Protoplasmic expansions sometimes sprout in all directions, and the corresponding corpuscles show rounded outlines that are hardly interrupted by narrow ridges (Figure 4.1A, B, E, and F). Sometimes, however, there is a marked somatic bipolarity that corresponds to the emission of appendages in two main directions (Figure 4.1D and H). At other times, all the protoplasm is directed in one direction, forming one or two thick processes that soon divide into secondary and tertiary branches (Figure 4.1I). However, as to the fundamental pyriform, spindle, round, and polyhedral types resulting from the presence or absence of protoplasm polarization are added the size variations and deformations due to the interposition of tissue elements (especially nerve fibers), from which a morphological spectrum arises beginning with small round and star-shaped cells, to continue with medium and large elements polarized in one or more directions, and ending in very long corpuscles where poor expansions are added to the soma stretching.

## Processes

Oligodendrocyte processes show different aspects, depending on circumstances. Thus, the silver carbonate technique sometimes presents them as very thin threads and sometimes as dim tracts in which the double contour is perceived well. The modified Golgi method also produces similar aspects (Figures 4.2, 4.4, 6.3, and 6.10); leaving strip forms visible, especially in the thick protoplasmic appendages (Figures 4.5, 6.5, and 9.9D). The impression on examining the most successful slides is that expansional protoplasm is formed by wider or narrower very subtle tracts which due to reagents, bend, twist, and narrow, becoming filiform in appearance, that cannot be confused with Ranvier–Weigert fibers.

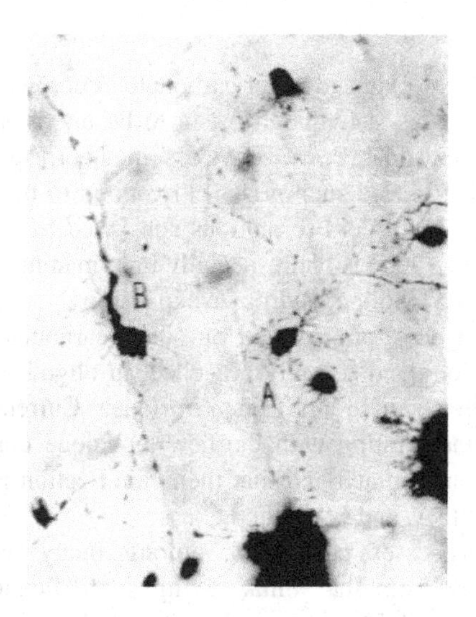


**Figure 4.2** Oligodendrocytes of the first (A) and second type (B) with filiform and nodose expansions. Cerebral white matter of the cat (Microphot.).

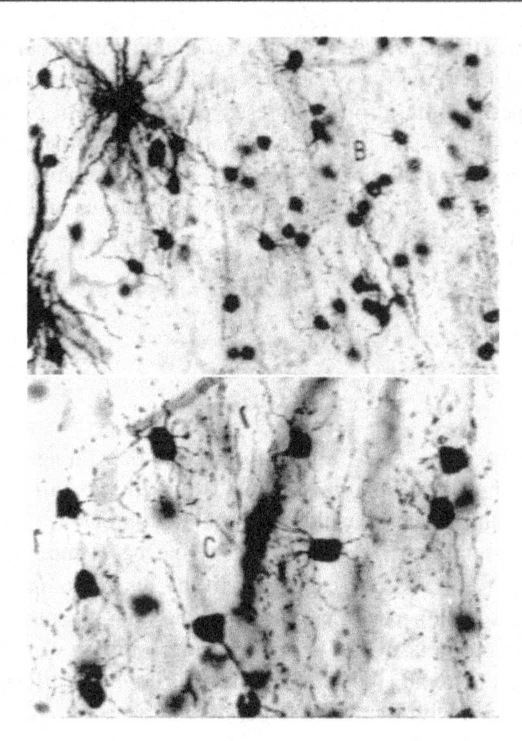


**Figure 4.3** First type Oligodendrocytes in the cerebellar white matter of the cat, seen here with two different magnifications: A, fibrous astrocyte; B, Robertson's oligodendrocytes; C, oligodendrocytes with abundant nodose processes. (Microphots.).

In both the platinum chloride and silver carbonate or chromate stainings, the regular line of cellular processes is frequently seen to be interrupted by small thickenings or nodes (Figures 4.2 and 4.3), which we assumed corresponded to dichotomies. Recent studies show that indeed such nodes correspond to true dichotomies, collaterals, or simply to plate or discoid formations supported by nerve fibers. However, ampullar thickenings are often visible (especially in human material) that correspond to nerve tubes unsheathed by oligodendrocyte expansions.

On making our first description based on silver carbonate stainings, we noted the lack of processes, long and scarcely branched, in oligodendrocytes. These facts were then confirmed by Penfield and López Enríquez. Currently, our approach has evolved a little on demonstrating with the new technique that the number of cell processes is greater than assumed and that their ramification is often very complex (Figures 4.4A and B, 6.11A, and 6.12A and C).

Oligodendrocyte processes can show, among many others, the following aspects: radiation away from the soma, giving some flexuous course branches (Figures 4.7 and 6.14A and C) that are often oriented in a predominant direction (that of the nerve fascicles) (Figures 6.11, 7.4, and 7.5); after a variable course,

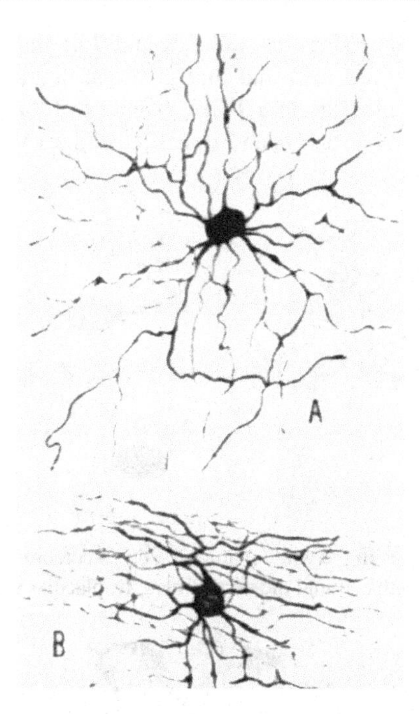


**Figure 4.4** Oligodendrocytes with many filiform and dichotomized processes that radiate in all directions (A) and mostly oriented in the direction of nerve fibers (B). Cerebral white matter of the cat.

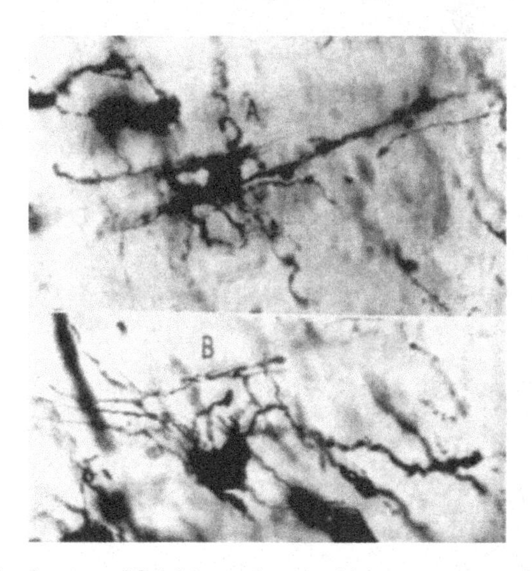


**Figure 4.5** Oligodendrocytes of Cajal (second type) with strap processes that in A are spirals, surrounding myelin tubes, and in B are divided in T following the nerve fibers; canine cerebrum (Microphots.).

they approach a nerve fiber (Figures 4.6, 7.3, and 7.8) and accompany it, closely united with it, until they are lost; and that give one or more branches at almost right angles that end up dividing in a T-like manner to follow nerve fibers in their two directions (Figures 4.7A, 6.11A and B, 6.12C, and 6.13).

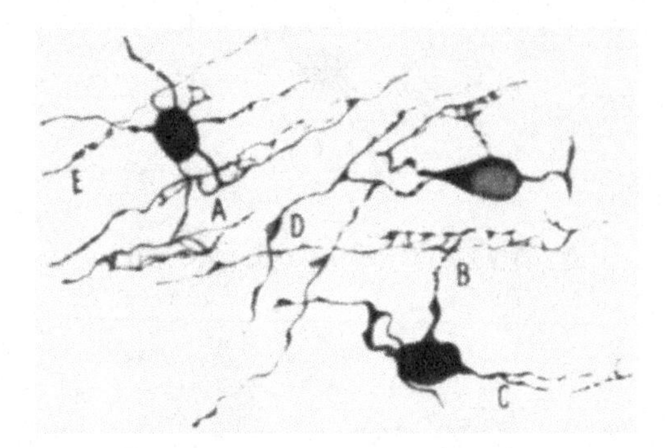


**Figure 4.6** Oligodendrocytes in the white matter of a dog's cerebrum whose extensions A, B, C, and E form reticular sheaths around the nerve fibers; D, placular widening.

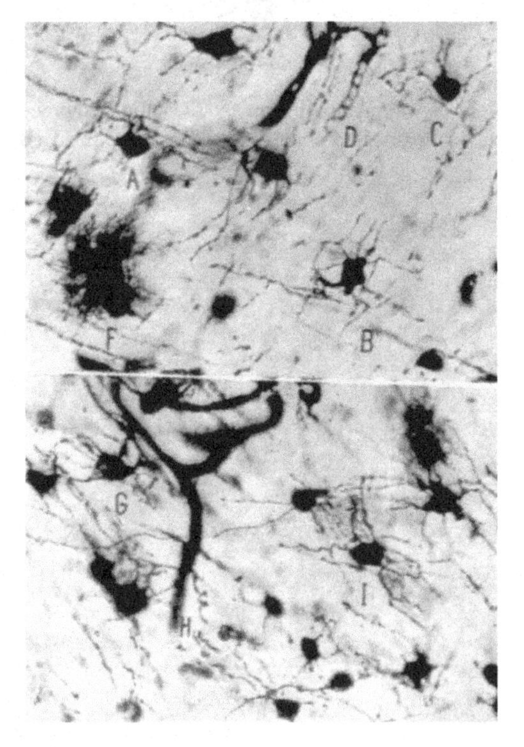


**Figure 4.7** Two fields of cerebral white matter of a dog with oligodendrocytes of the first and second types, whose expansions divided in T are shown in A, B, C, G, and I accompanying the nerve fibers; D, forming reticular extensions around a medullated axon; E, couple of dwarf astrocytes; H, vessel (Microphot.).

## Connections with Nerve Fibers and Vessels

The careful study of cytoplasmic expansion behavior reveals their general tendency to approach the nerve tubes and follow them for very long distances and create very faint sheaths around them. Undoubtedly, there are cell processes whose intimate association with nerve fibers cannot be seen and appear to be disoriented in the complex web of tissue; but branches of unknown origin that end up widening on medullated tubes are frequently observed. This makes the hypothesis that oligodendrocytes lack freely completed expansions and that they all finish on the myelinated surface of the nerve fibers very plausible. This assumption has so many possibilities for us to confirm that we have based on it the morphological and physiological concept of the oligodendritic cell web.

Contrary to Bailey and Schaltenbrand and Jakob's supposition, describing vascular feet on the oligodendroglial extensions, these show no tendency at all to implant themselves in vessels. Furthermore with silver carbonate and chromate dyes, which are the most selective that might currently serve us, it is impossible to demonstrate the existence of a single pedicle widened over vascular surfaces except for those from real astrocytes. Robertson and Perugini already noted the lack of relations with vessels in mesoglia cells and spinal cord small gliocytes, respectively. Penfield, López Enríquez, and Marchesani find an essential difference from astrocytes in the absence of implants or vascular pedicles in oligodendroglia. These are found attached to nerve fibers, and if they have any relationship with vessels, it is merely contact, when their processes pass the wall tangentially or continue along this, linked to nerve fibers (Figures 4.8 and 4.9).

Nevertheless, whereas each oligodendrocyte has purely eventual mediate relations of little interest with vessels the overall view of oligodendroglia offers perivascular dispositions, the significance of which we will discuss later.

## Expansional Plexus

The oligodendroglia are made up of very abundant cells living alone or in small groups in the gray parts (Figure 6.1), and they form true pleiads in the cerebral, cerebellar, and spinal cord white matter (Figure 4.10). Assuming that each oligodendrocyte emitted three to six dichotomized expansions, the plexus generated by them would be extremely complex. With the protoplasmic appendages more numerous than predicted and their branchings also more abundant (Figure 4.4A and B), the complication of the web that forms is beyond what could be suspected.

The primary and secondary expansions that flexuously run in different directions longer or shorter spaces before approaching and follow the corresponding myelin sheaths, together with the peritubular formations (plates, bridles, infundibula, rings, and lattices), breed such a tight plexus that Alzheimer's school researchers would describe and qualify it as a syncytium. Indeed, if there is a neuroglial syncytium, the main part would correspond undoubtedly to the oligodendroglia, which, if such an

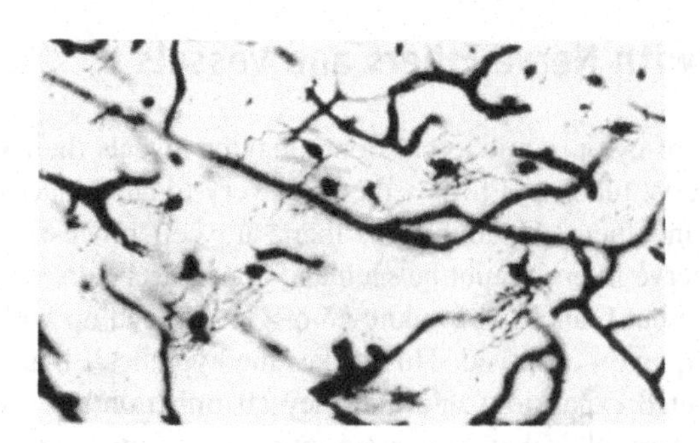

**Figure 4.8** Oligodendrocytes of the first and second types in the cerebellar white matter of a cat. The expansions have no contacts with the vessels (Microphot.).

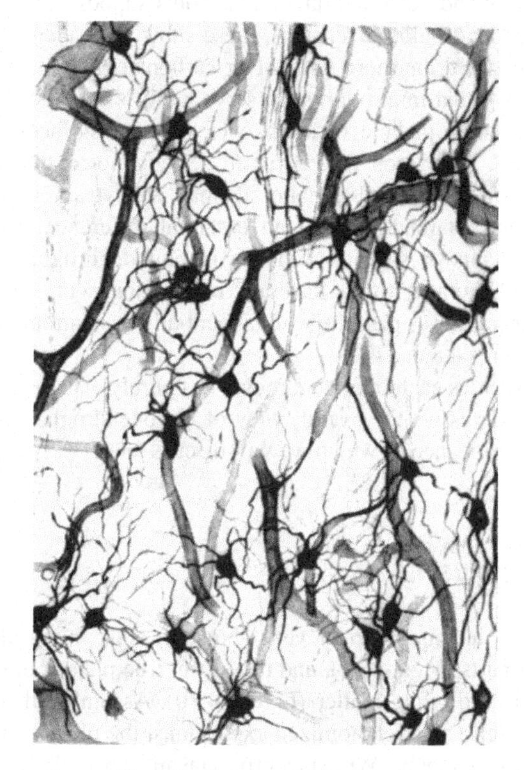

**Figure 4.9** Small vascular satellite oligodendrocytes in the white matter of the dog's cerebellum. There are no real vascular connections.

approach were accepted, would form a continuous three-dimensional reticulum in which mainly nerve fibers and vessels would be submerged. In our opinion, however, there is no such a reticulum and although in some stainings, the dye deposited at the crossing points of the cytoplasmic expansions mimics soldering and syncytial appearances, which is closest to reality, as we see it, is that the ramifications of

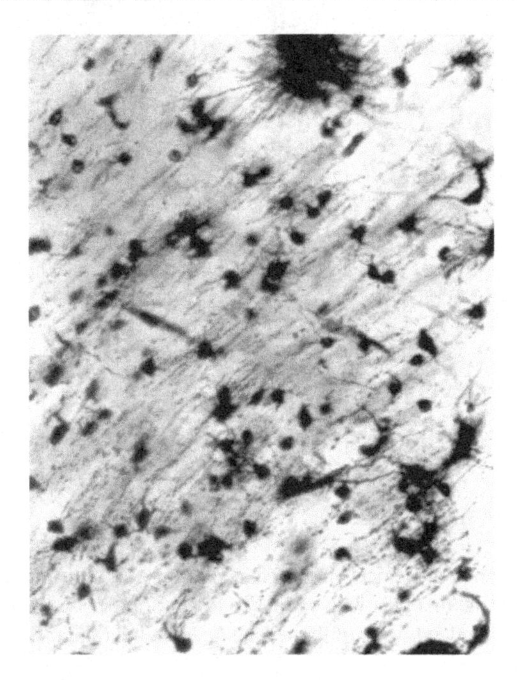

**Figure 4.10** Set of oligodendrocytes in the cerebral white matter of dog. The expansions are nodose and tend to be oriented in the direction of nerve fibers (Microphot.).

oligodendrocytes engender a diffuse plexus (never a mesh with nodal points) surrounding nerve fibers and capillaries (Figures 4.12–4.18). In addition to that plexus, as soon as cell expansion touches a nerve fiber, it widens over it forming polymorphic sheaths where small lateral branches abound or there are reticular or fenestrated aspects and infundibuliform or ring-shaped strengthening (Figures 7.11A and E and 7.13A–C).

Bailey and Schaltenbrand had thought about oligodendroglial participation in the limiting perivascular membranes, but this hypothesis has not been based on objective observations for this simple reason. The techniques used to study the *membranae limitantes gliae* do not stain oligodendrocyte processes; so, there is no doubt that these belong to them and not to other neuroglial types. Without now going into the bottom of the problem, so much worked already about the vessel marginal membranes and pia mater (problem that could be discussed widely), we should not omit our conviction that the limiting barriers between mesodermal and ectodermal elements have different characters than usually assigned. There must be differentiated in them the share corresponding to astrocyte insertions, whose feet are implanted in the perivascular connective tissue and that corresponding to the diffuse plexus of the neuroglial expansions traversed by the vessels (Figures 4.15, 4.16, and 4.18).

In the gray matter, the neuroglial web is mainly formed by protoplasmic astrocytes that constitutes Cajal's polygenic plexus (equivalent to the diffuse three-dimensional network of the German school), which is closely related to nerve cells, ensheathing their dendrites and forming to their bodies a kind of discontinuous coats,

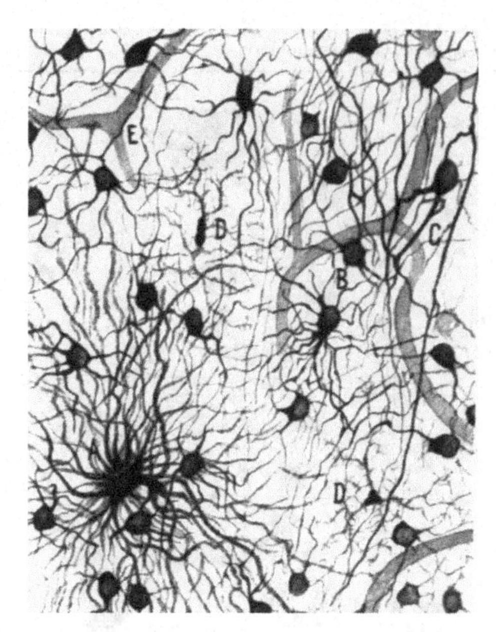


**Figure 4.11** Set of interstitial elements in the cat cerebrum A, fibrous astrocyte; B, oligodendrocytes of the first type with variable number of processes, many of them divided in Y or T; C, oligodendrocyte of the second type; D, microglia; E, vessel.

**Figure 4.12** Set of oligodendrocytes in the cerebellar white matter of the cat. The protoplasmic processes form a rich plexus that envelops nerve fibers and vessels.

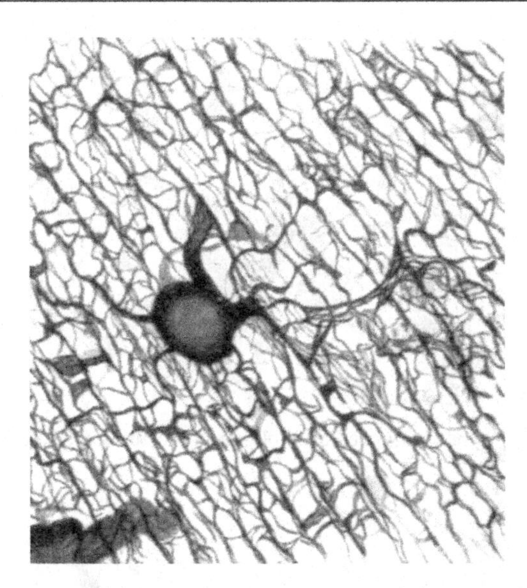

**Figure 4.13** Detail of the plexus formed by the oligodendroglia in the cerebellum of the cat. The cell processes and their branches tend to follow the guidance of nerve fibers and form complex ensheathings.

some bulky elements of which have been described with the name *Golgi's network*, and in whose genesis the oligodendroglia barely seems to intervene, according to all indications (Figure 6.1).

The main web of the white matter is constituted by the expansions of the oligodendroglia, which are intimately connected with nerve fibers to which they form special sheaths.

Fibrous astrocytes (Figures 4.11A, 4.12, and 6.10G), although participating in the general web and spreading all over their very long processes, only show close relationship with the pia mater (astrocytes of the cerebral molecular layer, Bergmann's cells, and so on) and with vessels, but not precisely as a means of separation.

It is, therefore, that as protoplasmic astrocytes are attached to nerve cells, oligodendrocytes fulfill their mission close to the nerve fibers, unsheathing them, whether or not surrounded by myelin, and fibrous astrocytes act through pial and vascular connections.

The general plexus of astroglia and oligodendroglia surrounds the vessels entirely but does not reach condensation around itself to form insulating lamellae; instead, it retains the disposition corresponding to a simple fibril crossing over. Only the widened feet of fibrous and protoplasmic astrocytes (the latter to a lesser extent) would be able to form the vascular limiting membrane.

Regarding the neuroglial syncytium we have rooted convictions contrary to its existence, although the plexus of neuroglial expansions we imagine is so broad and general and made of meshes so narrow that discrepancies in the morphological interpretation of syncytium (protoplasm fusion) and plexus (expansion interlaces) in no way affect the functional concept in relation to the nervous elements. In order for

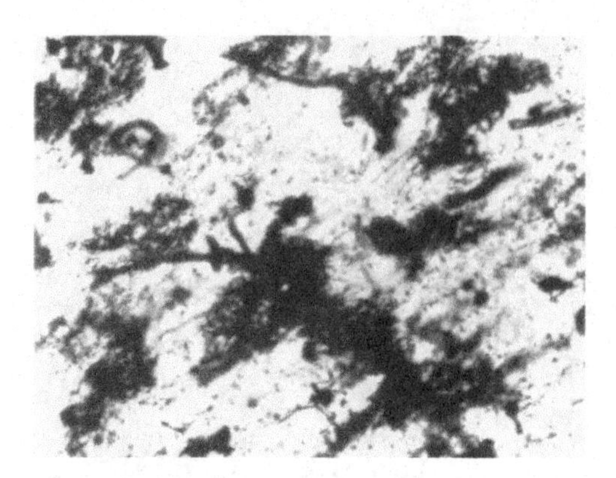

**Figure 4.14** Plexuses formed by processes of oligodendrocytes in the cerebral white matter of cat (Microphot.).

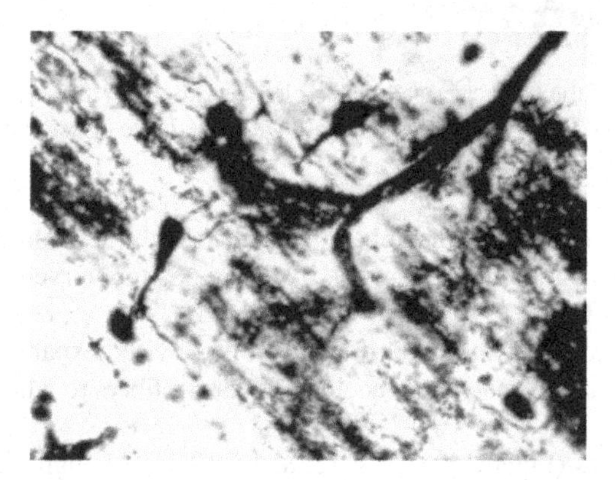

**Figure 4.15** Detail of the expansional plexus of oligodendroglia with the participation of some elements of the second type (Microphot.).

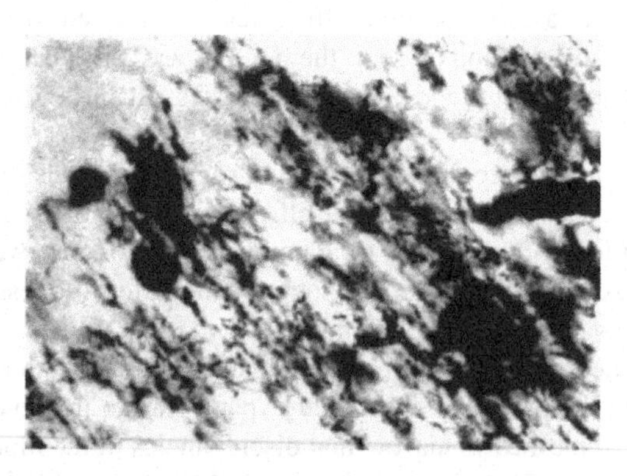

**Figure 4.16** Syncytial appearance of the expansional plexus of oligodendrocytes in the cerebellum of a cat (Microphot.).

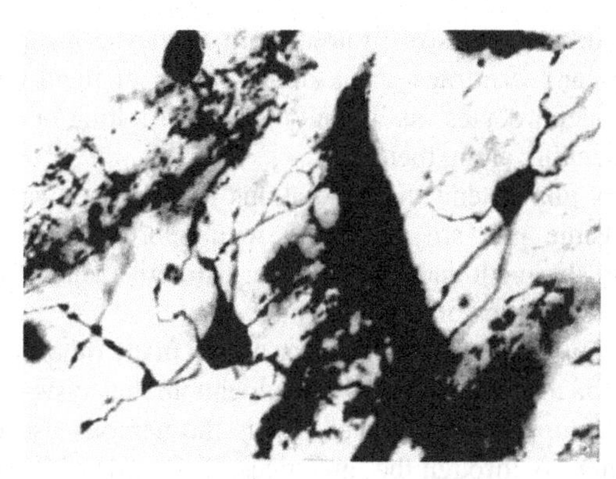

**Figure 4.17** Participation of oligodendrocytes in the perivascular neuroglial plexuses, cat cerebrum (Microphot.).

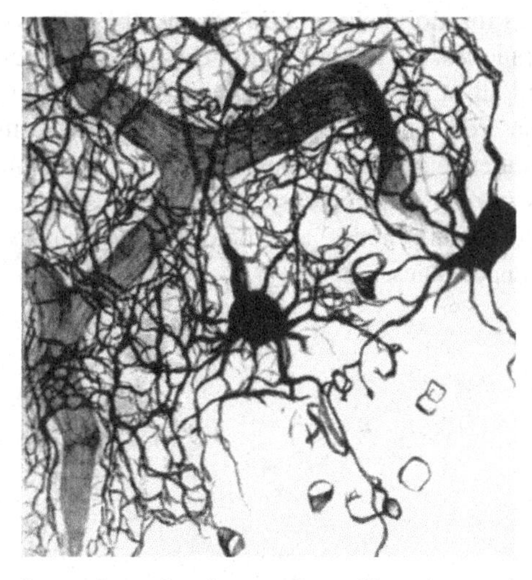

**Figure 4.18** Very complex perivascular plexuses formed by crisscrossing expansions of oligodendrocytes.

them to be separated and protected by the interstitial web, so that the neuroglia can exercise a trophic or antitoxic function and elaborate secretory products, their originally isolated elements are not required to merge. In the case of oligodendroglia, we can say that their cytoplasmic expansions do not anastomose with each other or with those belonging to astrocytes.

Although we dislike making conjectures without solid foundation based on the observation of undisputed phenomena we must do so in relation to the architectural constitution of the neuroglial structure.

According to this conjecture, the neuroglial ramifications are not anastomosed to form syncytium, but nor do they end freely, lost in the nerve structure. The fibrous astrocyte processes, either long or short, always end implanting on the vessels or on the pia mater. The arborizations of protoplasmic astrocytes mostly end up leaning on neuronal somata and dendrites and a minimal part of them on capillaries. The appendages of oligodendrocytes finally end their course, long or short, widening on nerve fibers and extending along them.

Thus, the fibrous glia extend their radiations among their vascular or pial surfaces; the protoplasmic glia are associated with neuronal structures, and oligodendroglia surround the medullated tubes. The distinctive role of each of the three neuroglial varieties is thus anatomically outlined.

As for microglia (Figure 4.11D), which for us and many other researchers are outside the true neuroglia and belong to the reticuloendothelial system, they remain free in the complex of cytoplasmic processes that is the nervous tissue extending their appendages like tentacles through the interstices of the tissue. Bergmann's idea that microglia also participate in the syncytium and have morphological transitions with oligodendroglia is beyond our comprehension capability. Indeed, we cannot accept that among genetically different elements relationships other than one of mere contiguity can exist. And as for morphological transitions, they are evidenced only by the use of imperfect techniques. As the microglial elements and many of the oligodendroglia are small, and all of them with prolongations, confusion is easy when dealing with round nucleus cells and incompletely stained protoplasmic appendages and without their real features; e.g., zigzag course, abundant dichotomies, terminal tufts, and collateral spines in microglial processes; undulating course, few divisions, frequently T-shaped, occasionally nodes, and a close relationship with nerve fibers, in the oligodendroglial appendages.

# 5 Textural Characteristics of Protoplasm in the Oligodendrocytes

The general texture of the cytoplasm does not differ in some points from that of the astrocytes, but in other cases they fundamentally differ. Current technical resources do not allow determination of the actual constitution of oligodendroglia protoplasm, which is almost completely resistant to staining with dyes. With metal impregnations, it usually occurs stained in blackish color without showing any organization. However, with favorable stainings obtained with silver carbonate, which is the most appropriate of all metal reagents to reveal structures, one can see some details that have the appearance of corresponding to the true architecture.

## Spongioplasm

The somatic protoplasm of oligodendrocytes is very delicate and has a fine lattice, spongy appearance, whose exaggeration, in certain stains where the corpuscles are swollen and almost vesicular, shows it is an easily vulnerable cytoplasm (Figures 8.6, 8.7, 8.12, and 9.1). The frequent spheroidal configuration of corpuscles demonstrates the existence of a marginal condensation of the cytoplasm, which undoubtedly forms an imperceptible cover.

Just as in astrocytes, in oligodendroglia specific granules are housed in the lattice holes. In general, the structure more or less resembles protoplasmic glia (depending on whether the staining is strong or weak).

In contrast to the fibrous astrocytes, whose protoplasm is traversed by fibers that often show expansions and to protoplasmic astrocytes that unusually can originate fibers by differentiation of spongioplasm, oligodendrocytes in neither this normal state or under pathological conditions exhibit a true fibrillar structure. Jakob's observations in this regard would require a review by less fallible techniques than the current. The expansions of the oligodendritic corpuscles are comparable to more or less wide ribbons (Figures 4.5, 7.3, 7.8, 7.11–7.13) that are modeled on nerve fibers often offering fenestrated and torn aspects because of excessive strain and breakage of the fine meshes of their spongioplasm. However, such expansions, because of a phenomenon of transverse retraction can acquire (and frequently show) the appearance of real fiber, not just on their way from the cell body to the corresponding nerve fibers, but also along these fibers (Figures 6.2, 6.3, 6.10, 6.11A). Protoplasmic retraction thus makes clear the annular and spiral dispositions of the processes around the nerve tubes (Figures 7.11–7.14) and laminar, reticular, and infundibular

**Río-Hortega's Third Contribution to the Morphological Knowledge and Functional Interpretation of the Oligodendroglia.**
**DOI: http://dx.doi.org/10.1016/B978-0-12-411617-7.00005-5**
© 2013 Elsevier Inc. All rights reserved.

condensations and reinforcements, which are closely related to nerve tube myelin sheath (Figures 8.1, 8.2, and 8.4).

In some wide rings (Figures 7.19H and 8.1D), and especially in the infundibula and diaphragms dividing the myelin (Figures 7.19I, 8.4A, and 8.5), passing from its cover to the axis cylinder, there is a beginning of fibrillar differentiation manifested by very fine lines, parallel or circularly interweaved, that resemble baskets.

It is possible to identify the amorphous or poorly differentiated part of the cytoplasm in oligodendrocytes, in addition to the granular structure corresponding to all cell types (represented by the chondrioma), the centrosome and Golgi apparatus. These organelles have already received special descriptions, and need only be remembered to complete the examination of the protoplasm intimate texture.

The existence of rudimentary Golgi apparatus, comprising little clots and cords placed on one side of the nucleus, was described by Cajal in 1914. The bicentriolar centrosome located in the vicinity of the nucleus was pointed out by us in 1921. Finally, the chondrioma completing the trophic organization of oligodendrocytes also was described by us in 1925.

The oligodendroglia, on the other hand, are probably responsible for two important functions: (1) purely passive protection and isolation, unsheathing and separating the nerve fibers and tubes, which corresponds to amorphous protoplasm; and (2) active production of specific substances serving, by all indications, the formation of myelin. In oligodendrocytes there is, therefore, a secretory function corresponding to the neuroglia overall activity that is manifest by the presence of gliosomes in spongioplasm hollows.

## Gliosomes

Nageotte's, Mawas's, Achúcarro's, and Cajal's studies among others lent great importance to the neuroglia granular structures as corresponded. Therefore, we need not insist on this point but fully accept the functional concept they evoked after verification of cytological data supporting this concept.

Studying, in separate publications, the chondrioma and specific granules of neuroglial cells, using selective variants of silver carbonate technique, we described oligodendroglia gliosomes, extending the 1921 description. In the somatic and expansional protoplasm of oligodendrocytes are granules with shifting thicknesses that appear extraordinarily abundant during certain times of life.

In young animals, when the myelination of nerve fibers is more active, the protoplasm of oligodendroglia elements contains very abundant secretory granules. Penfield's studies (1926), corroborating ours, demonstrate not only the reality of specific granule formation but also that its maximum takes place early in life.

We cannot go into detail about the possible relationship between oligodendroglia secretory activity and myelination, being a theme with ongoing experiments and about which we still lack a definite opinion (Figure 9.10). We may return to this important topic at some later date to deal with it in detail. For the moment, suffice it to state the working hypothesis and to point out the morphological data that suggest it.

# 6 Morphological Varieties of Oligodendrocytes

Although we tried in the preceding chapter to describe the general characteristics of oligodendrocytes, we are not sure we achieved our purpose. The variety of aspects oligodendroglial elements offer is undoubtedly so large, that if each type did not constitute a step in the uninterrupted morphological series, one would be forced to hesitate prior to identifying the large and small elements with polymorphic expansions.

Our study would be incomplete if we did not report the different shapes and sizes and the progressive differentiation of myelin sheaths. Because it is difficult to give each type an appropriate name according to their characteristics, we prefer to denominate them, in short, following a sequential numbering which fits in part to variations in size, but in tribute to Robertson, Cajal and Paladino, we have used their names for as many types of oligodendrocytes.

## First Type

Being probably the only one observed by Robertson, we consider it just to award the name of the Scottish author to this type.

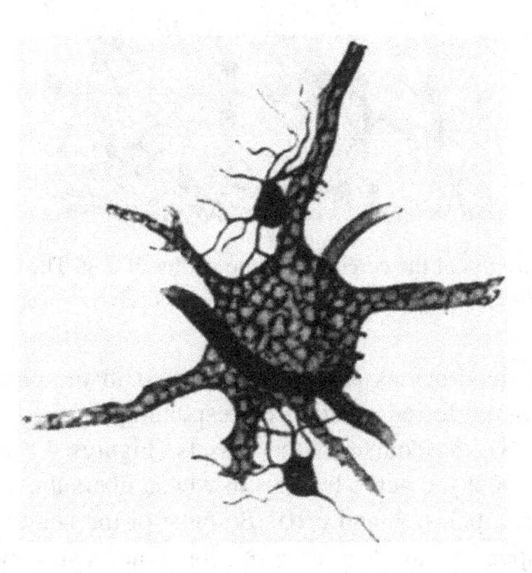

**Figure 6.1** Nerve cell with Golgi network, in which oligodendroglia have little participation.

Río-Hortega's Third Contribution to the Morphological Knowledge and Functional Interpretation of the Oligodendroglia.
DOI: http://dx.doi.org/10.1016/B978-0-12-411617-7.00006-7
© 2013 Elsevier Inc. All rights reserved.

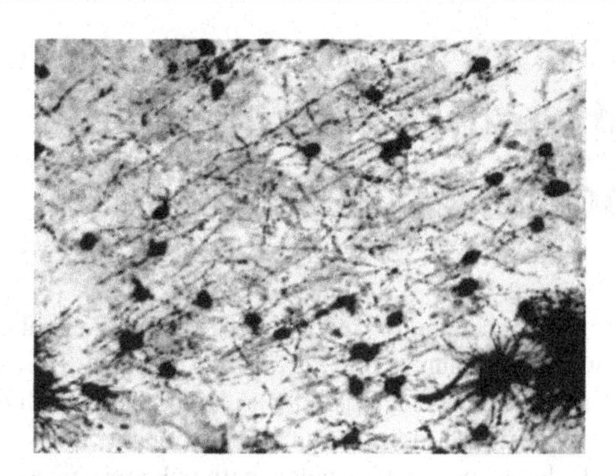

**Figure 6.2** Oligodendrocytes of the cerebellar white matter of cat, whose expansions, long and nodose, are oriented along the medullated fibers (Microphot).

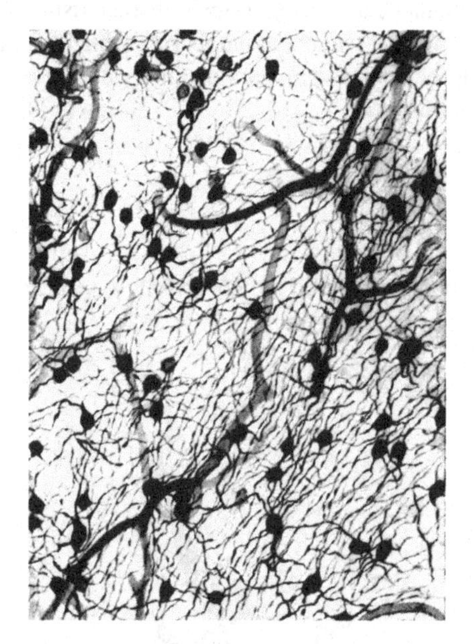

**Figure 6.3** Oligodendroglia of the cerebellar white matter of dog. The cells show very long knotty processes like fibers, interwoven in a simple plexus.

Robertson's oligodendrocytes are found scattered in the cerebrum, cerebellum, and spinal cord, their preferred position corresponding to the contour of the nerve cells (Figures 6.6–6.8), the course of the vessels (Figures 4.8 and 4.9), and especially to the interstices of the nerve bundles in whose fibers the myelin sheath is thin (Figures 4.3, 4.10, 6.2, 6.3, 6.9, and 6.10). So most of the neural and vascular satellite elements correspond to the first type of oligodendrocytes, whose characteristic clusters can be observed in the gray parts of the cerebrum, cerebellum, and spinal cord and the white parts of the cerebral gyri. Almost every element lying adjacent

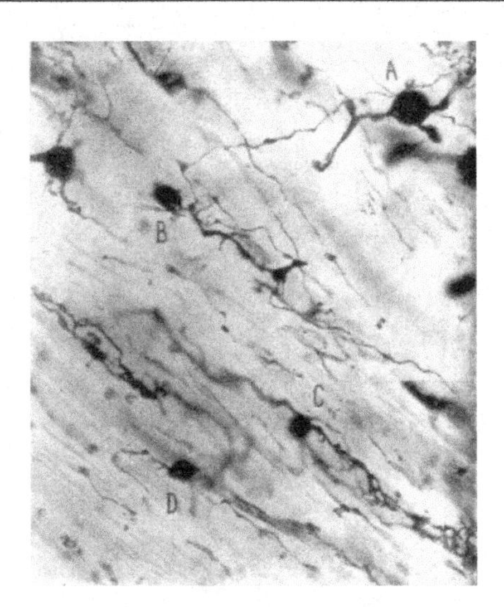


**Figure 6.4** Oligodendrocytes that form reticulated sheaths around the nerve fibers. A, second type element; B, C, and D, elements of the third type. Cat cerebellar white matter (two joined microphotographs).

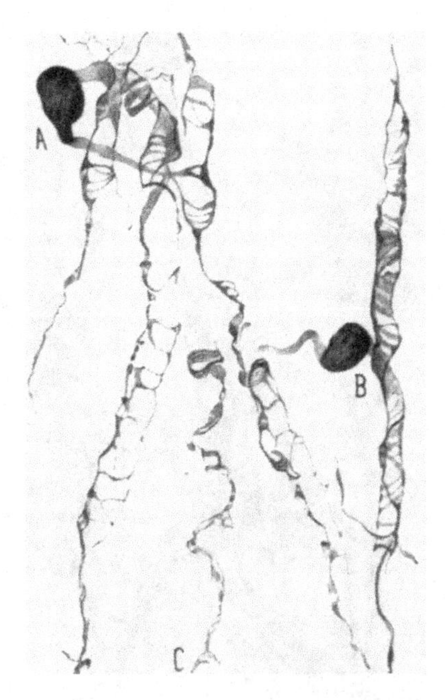


**Figure 6.5** Oligodendrocytes of the third type in cat's medulla. A, element whose processes form reticular sheaths for three nerve tubes; B, element with an enlarged laminar process around a medullated tube; C, spiral expansion of an oligodendrocyte ending in a ring.

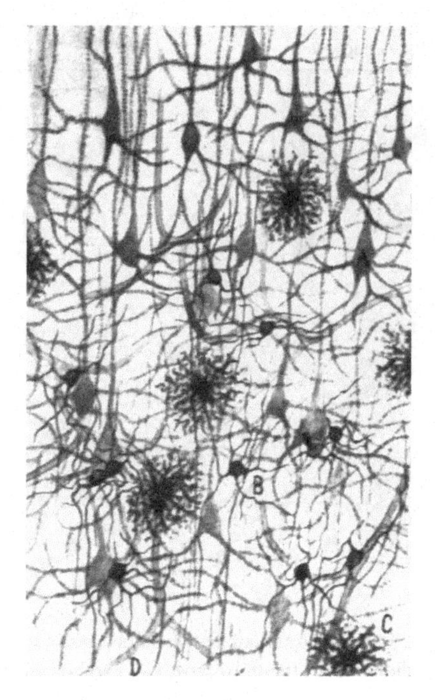


**Figure 6.6** Appearance of the cerebral cortex of cat with neurons (A), oligodendroglia (B), protoplasmic glia (C), and vessels (D).

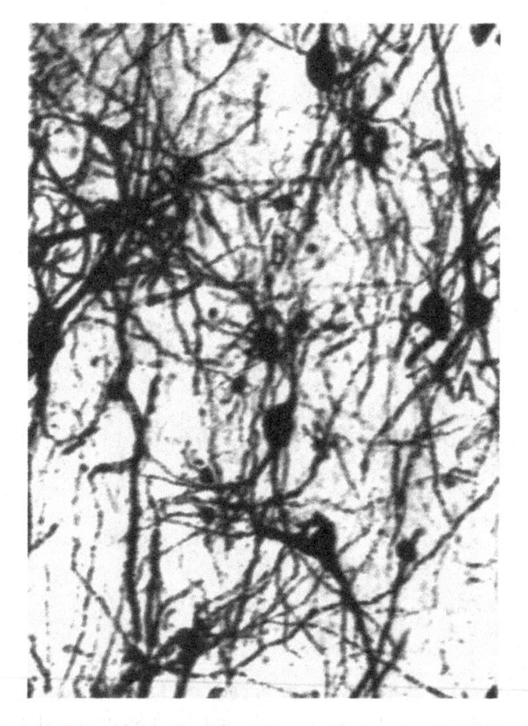


**Figure 6.7** Appearance of the cerebral cortex of cat with nerve cells (A) and oligodendroglia (B) (Phot).

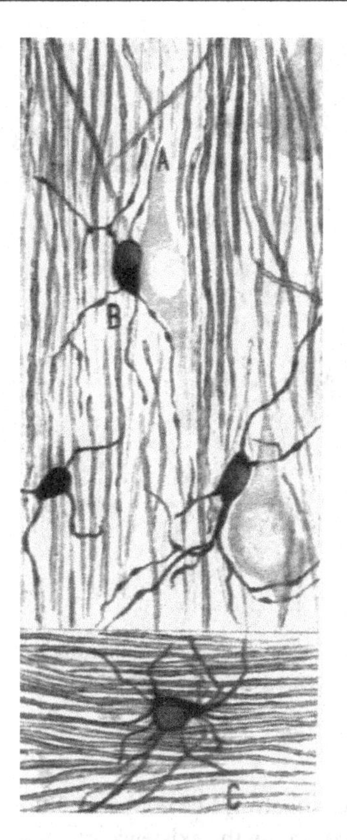


**Figure 6.8** Oligodendrocytes of the normal human cerebrum; their processes ensheathe thin nerve fibers. A, pyramidal cell; B, satellite oligodendrocyte; C, interfascicular oligodendrocyte in the white matter (silver carbonate stain).

to nerve cells belong to Robertson's oligodendrocytes; close to vessels, they are interspersed among corpuscles of the second and third types, and they are much less abundant among the nerve bundles. Since the study of topographic variations of distribution can only be carried out after careful examination we will leave our final opinion hanging in the air for now, yet solely by way of indication we provide the preceding data, collected from preparations which like all those of the Golgi method are incomplete.

The elements of the first type (Figures 6.9 and 6.10) are characterized by their small size (15–20 μm), rounded or polyhedral soma outline, and high number of very fine expansions emerging, stemming almost brusquely, widening just at their origin and the soma without losing their warped appearance.

With the modified Golgi method, they appear diffusely stained blackish; with some frequency, the nucleus location is denoted by a clear space that is almost always eccentric.

The processes arising from the perinuclear protoplasm appear in varying numbers, and although it cannot be known accurately if they are more scarce in some elements

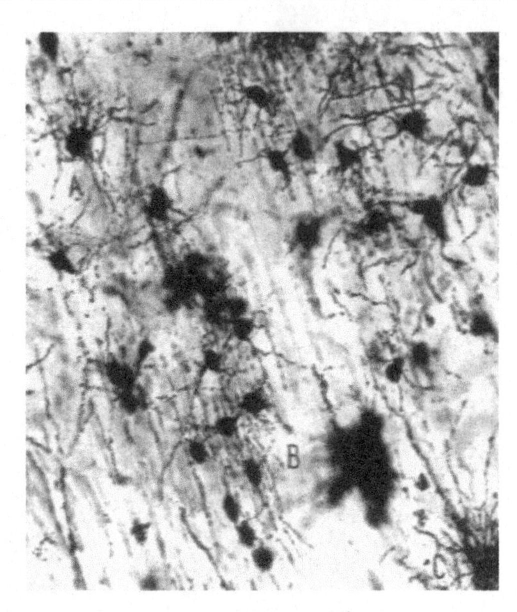


**Figure 6.9** Oligodendroglia of the cerebral white matter of dog: A, element with multiple processes; B, element whose appendages follow the direction of the nerve fibers; C, fibrous astrocyte (Microphot).

than in others because there will always be doubt about the complete staining of some of them, it can be considered certain the existence of corpuscles with abundant expansions (15, 20, or more) along with others with smaller numbers of them (5–10).

From the way cell appendages arise and orientate several morphological forms can be recognized: (a) with processes sprouting from all around and radiating in multiple directions (Figures 6.9A, 6.10A and B, 6.11A and B, 6.12A and C, and 6.14A), (b) with processes preferably located at the poles of more or less elongated elements and directed from their origin parallel to nerve fibers (Figures 6.10D and 6.13), (c) with processes issued in small groups or tufts from thick somatic mammillary lumps (Figures 6.12A and 6.14C). In the observations at low magnification, when the thin processes are invisible, the preceding types appear rounded to ovoid or in two, three, or more protoplasmic protrusions. Monopolar aspects are not uncommon.

The expansions form small rounded cords at their origin, but, after division, as soon as they usually approach the nerve fiber surface (myelinated or not) they widen to wrap them. The fate of cytoplasmic appendages is that, which they all fulfill, so each oligodendrocyte gets almost as many relationships with nerve fibers as it does processes. The division of the processes is usually very restrained, the visible dichotomies not exceeding two to four. These take place in very wide angles, the most common ending being in a T-like manner, and each of these corresponding to a nerve fiber.

Although the shrinkage suffered by the expansions surrounding nerve fibers makes them appear as more or less nodose threads (Figures 6.11–6.14), sometimes

**Figure 6.10** Neuroglial elements of the cerebral white matter of cat: A, oligodendrocytes with long expansions that follow nerve fibers; B–D, oligodendrocytes with numerous processes divided in T; E and F, dwarf astrocytes; G, fibrous astrocyte.

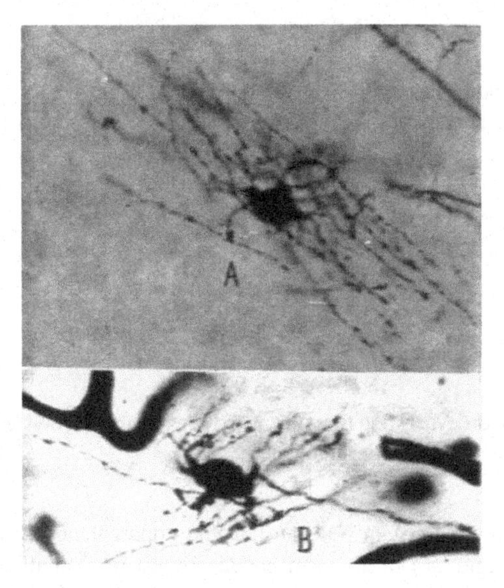


**Figure 6.11** Two oligodendrocytes of the first type with abundant nodose processes, some divided in T (A and B) (Microphots).

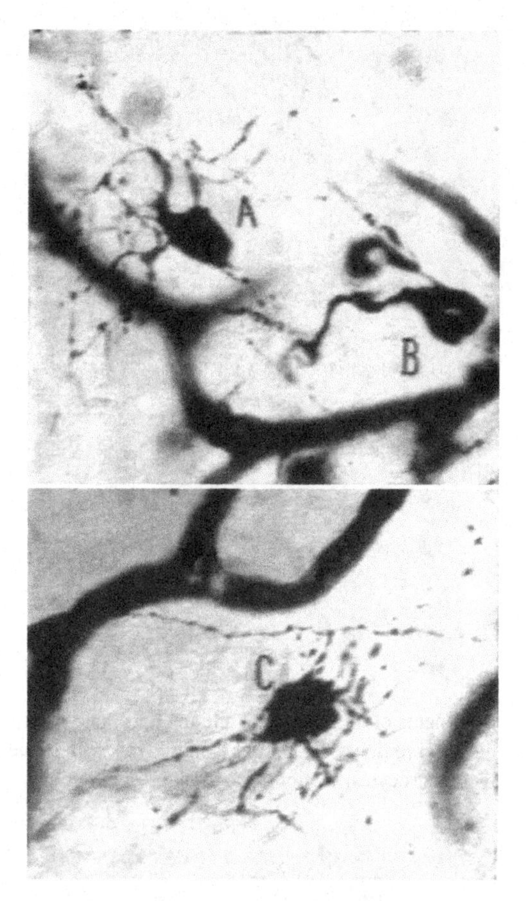


**Figure 6.12** Oligodendrocytes of the first type next to vessels: A, with a tuft of expansions; B, with a single stained process C, with multiple appendages divided in T (Microphots).

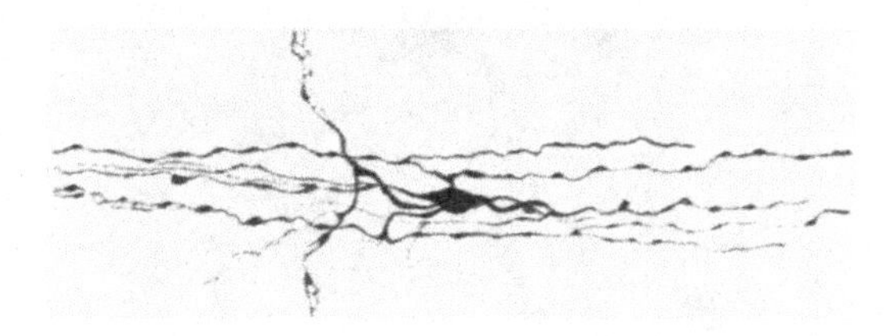

**Figure 6.13** Oligodendrocyte of the first type with long expansions divided in T, with perimyelinic small plaques and delicate reticula.

with double neat contour, it can be seen that the apparent nodes, if not corresponding to dichotomy angle, are protoplasm condensations in a small plaque or fine ring manner. These formations, to which delicate reticular aspects are added, are much more developed when they correspond to thick tubes and in the expansions of bulkier

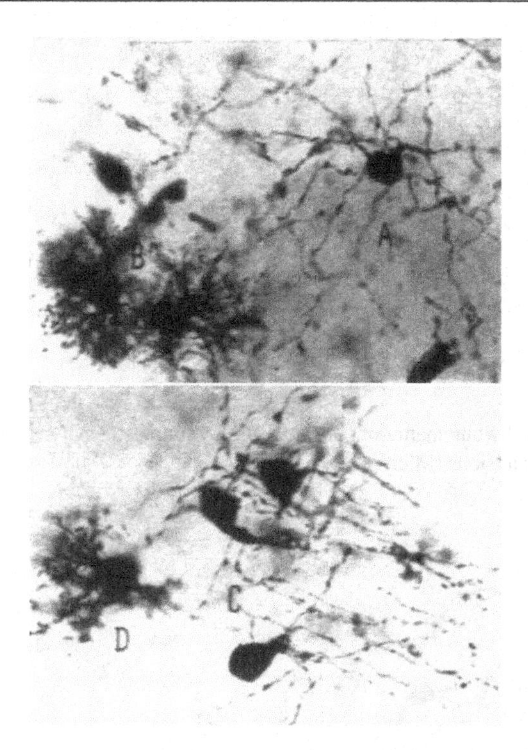


**Figure 6.14** Oligodendrocytes of the first type in cerebral white matter of dog: A, element with irradiated processes; B and D, dwarf astrocytes; C, oligodendrocytes with intertwined expansions (Microphot).

oligodendrocytes (Figure 4.6). Regarding the rings, with certain not very successful stainings for the original elements, they appear to be extremely abundant (Figure 6.15).

## Second Type

The elements of this second type not only differ from the preceding ones by the size, which reaches 20–40 μm or more but also differ in how their processes stem and are laid out. They tend to be polygonal in shape, often cuboidal; monopolar and bipolar types are not rare. Under medium magnification, they often show the mammillated outline seen by Cajal in some thick corpuscles of the third element; therefore, as homage to our master, we want to name them *Cajal's oligodendrocytes*.

Cajal's oligodendrocytes are not visible in the gray matter and, on the contrary, abound in the white parts. Therefore, good places to observe them are the axial region of the cerebral and cerebellar gyri (Figures 4.12 and 6.16), the oval center of the hemispheres, peduncles, pons, medulla, and spinal cord funiculi.

It is not easy to know the respective proportion of the elements of the first and second types. Despite appearing in the same slides, in none of these does staining of

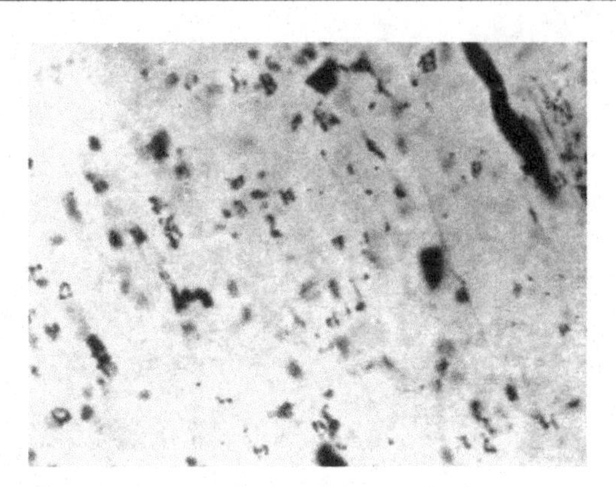

**Figure 6.15** Cerebral white matter of the cat with very abundant small rings in not stained oligodendroglial expansions (Microphot).

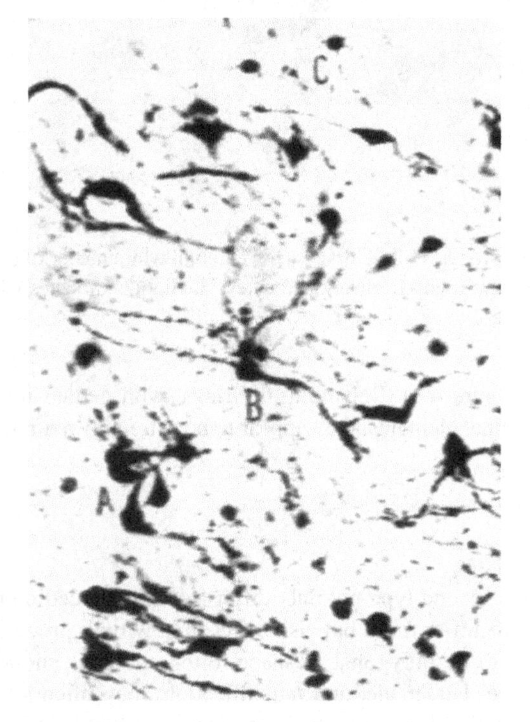


**Figure 6.16** Oligodendrocytes of the second type (A and B) in the cerebellar white matter of cat; C, element of the first type. (Along the corpuscle B are two elements of the first type.) (Microphot).

all interfascicular gliocytes succeed, although they often appear infinite in number. It is our impression, however, that most of the elements, which are arranged in interfascicular series, belong to the first type, although to all appearances, the thickness of the medullated tubes has great influence in their proportion. In this regard, it could

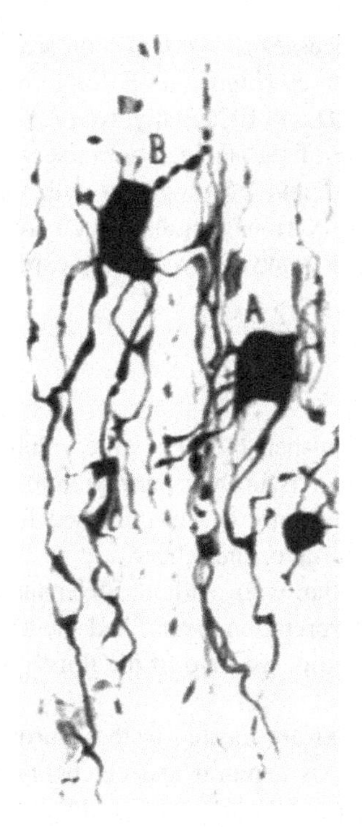


**Figure 6.17** Oligodendrocytes (A and B) of the second type in the cerebellum of cat, with tortuous expansions, rings, and reticula wrapping medullated tubes.

be ventured the rule, not well specified yet, that to thick nerve tubes corresponds a predominance of bulky oligodendrocytes, although in any case, types and sizes are extremely mixed, showing the gentlest transitions.

On examining the oligodendrocytes of the second type with immersion objective, one can see quite thick though not very abundant expansions from the somatic contour, which are then directed along the nerve fibers or are attached to them longitudinally after a short transversal distance (Figures 6.17–7.10). These expansions often follow a tortuous path changing the plane to curve on the medullated tubes, whose surface they run in a more-or-less spiral course, with circular turns and strap and reticular formations. Once attached to the myelin sheaths, they widen and simply condense at intervals in a ring shape, and small infundibula or are divided and subdivided on myelin drawing reticula.

The variety (a) appears poor in expansions, often from opposite poles, and frequently divided in a T-like manner (Figures 7.3B, 7.8A–C, and 7.10). Type (b) shows tuberous contours (Figures 4.5A and 7.6A and B) and stems a number of appendages, more or less dichotomized in a Y- or T-like manner. The form (c) exhibits one or more expanded arms extending away from the soma in various directions and stemming collateral twigs (Figure 7.7). Variety (d) is distinguished by the irregular, sometimes cuboid outline of the soma (Figures 6.4A, 6.17A and B, 7.1B, and

7.9A) from which several processes sprout that soon are directed along nerve fibers, ensheathing them in wide curves (Figure 6.4B) or forming reticula (Figure 6.4C), rings, and sleeves (Figure 7.1D and E). Finally, type (e) is characterized by the pyriform or fusiform appearance of the soma, transversely oriented to the medullated fibers (Figures 6.18, 7.1A, 7.4, and 7.5); together with some secondary appendages, they show one or two arms of various lengths, which are divided in a T-like manner or separated into a tuft of thick, wavy, and very long expansions.

## Third Type

Cells of this type are distinguished by their large volume, the particular aspect of their processes, and the characters in their perimyelinic sheaths. This variety of oligodendrocytes corresponds to many of the glimpsed formations, and attributed to astrocytes, by Paladino, Montesano, and others.

In homage to the wise Italian who, without understanding the precise characters and even having serious interpretation errors, had the shrewd inkling that a neuroglial skeleton existed for myelin, we give to the third type of oligodendrocytes the name of Paladino.

*Paladino's oligodendrocytes* are located in the neuroaxis territories, where there are thick myelin tubes, such as cerebral and cerebellar peduncles, pons, medulla, and spinal cord. They are especially abundant in the marginal fascicles of the latter organs mentioned.

Judging by what the best preparations show, the number of elements of the third type is much lower than in the two preceding. Our data regarding the exact share of the different varieties of oligodendrocytes are uncertain, but nonetheless we consider that an inverse relation exists between number and volume, not only in cells but also in the emitting processes. Contrary to protoplasmic and fibrous astrocytes that show crowds of appendages when large, these are scarce in large somatic mass oligodendrocytes attached to robust medullated tubes.

Cells of the third type appear truly oligodendritic, since they usually display one to four robust major expansions with some other flimsy and secondary ones. The predominant shape is, however, monopolar or bipolar (Figures 6.5, 7.11A and E, 7.12A and C, 7.13, 7.14, and 8.3A).

The bulky soma extends into one or two, rarely more, thick arms that follow a wavy course; once attached to their corresponding myelin tubes, they widen to cover them in large areas, not in a homogenous manner but by special arrangements in the manner of subtle laminae or reticula with a uniform thickness or with annular and infundibular reinforcements. It is more common that nerve tube contours and relief, that are colorless in the preparations (Figures 7.11 and 7.13), are recognized by extremely irregular sidebands of variable width, linked by cross and oblique bridles, sometimes very close, sometimes far apart, and always showing a fine striation that can be interpreted as incipient fibrillar differentiation of the cytoplasm. In some cases, the reticular aspect seems to correspond to fenestrations or tears, perhaps artificial, of the protoplasm stretched on the myelin sheath.

The peritubular expansions do not ordinarily follow a rectilinear direction over the myelin sheaths; but rather, they tend to make a spiral movement reminiscent of vine tendrils (Figures 6.5C, 7.11A, 7.13, and 8.3A). And like these, the coil spirals may be far apart or quite close, furthermore, it is not uncommon for them to form several rings together in some places. Thus, the nerve tubes being run by spiral bands with annular trabeculae and bridles, the appearances of the myelin sheaths are quite complicated. But there is even more complexity when a real protoplasmic ribbon division happens, and two or more spirals intersecting on the fibers (Figure 8.3A). Thus, as in the rare event that two or more expansions of different cells or the same cell coincide, there are very complexly organized myelin sheaths.

The arrangement adopted by neuroglial cover in thick tubes is extremely variable; special differentiations are distinguishable in it. Different appearances can be seen in the same fiber, whose polymorphism increases by the fact that numerous staggered oligodendrocytes are involved to make its protoplasmic case.

The spiroid, reticular, annular, and cylindrical formations initiate in the first-type element and appear well developed in those of the second and third types, where they are completed with infundibular formations or intramyelinic septa. In small- and medium-sized tubes ensheathed by Robertson's and Cajal's oligodendrocytes, fibers, lamellae, and superficial reticula with more-or-less numerous rings are visible, but these rarely possess well-formed funnels. Thick tubes ensheathed by Cajal's and Paladino's oligodendrocytes also have septa that interrupt the continuity of the myelin sheaths.

## Fourth Type

They are mono or very elongated bipolar elements that are highly reminiscent of Schwann cells. Although there are great morphological differences between them and the other varieties of oligodendrocytes, it is not difficult to discover the link by means of the corresponding transitions. These are corpuscles whose protoplasm, instead of forming appendages, was attached closely to nerve fibers, flattening on the myelin surface and providing superficial reinforcements to the medullary column (rings) and deep septa (infundibula).

The elements of this type are located in nerve bundles of peduncles, pons, medulla, and spinal cord, accompanying medium- and large-thickness medullated tubes.

They are characterized (Figures 7.15–7.18) by showing a fusiform soma, highly extended along the correspondent nerve fiber, whose surface it ensheathes by laminar, fenestrated, or reticular arrangements and with annular and conical reinforcements identical to those described in the preceding types.

In the Schwannoid corpuscles are two main forms: the bipolar, with more-or-less flattened soma, sometimes semicylindrical; and the monopolar, whose only visible process, transversely oriented to nerve fibers, bifurcates and becomes laminar as soon as it arises.

In fact, it is not always easy to accurately distinguish between the elements assigned to types three and four, since the shape is modified in them very little by little. There

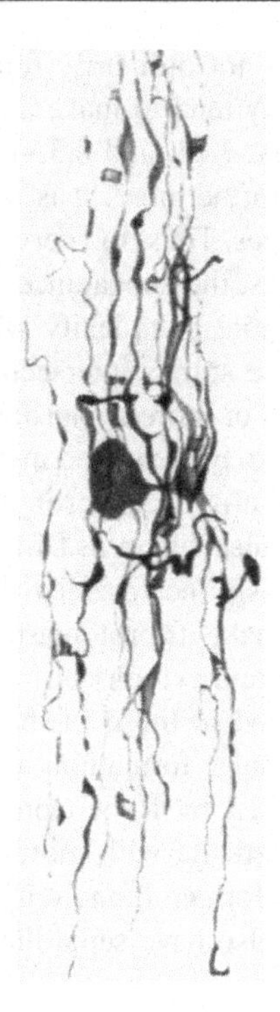

**Figure 6.18** Pyriform oligodendrocyte of the second type, with abundant processes accompanying and forming rings around nerve tubes.

is, moreover, an unbroken chain linking the smallest types of Robertson with the most branched and the Schwannoids through those of Cajal and Paladino. The classification made for purely descriptive purposes is therefore extremely artificial.

## Myelin Sheaths and Septa

This is in the overview (Chapter 4) where the expansions of oligodendrocytes, always in strips and frequently laminarly widened, to form real casings for the nerve fibers, thus one might think this neuroglial variety is morphologically differentiated to serve the myelin by way of support (and likely source of origin) and trophic vehicle of nerve fibers.

The characteristics of the myelin sheaths constituting Paladino's neuroglial skeleton are clearly not well-known today, because to interpret the images the best techniques offer us one needs to consider the possibility of them being artifacts. Discarding Paladino's and Montesano's observations with reference to the astrocytes (which in fact correspond to the structures of their extensions, mixed with those pertaining to the oligodendroglia) and although the methods of choice used make it difficult to incur in the mistake mentioned, misleading details may be provided which should be judged with caution. Thus, a definitive opinion cannot be maintained regarding perimyelinic membranes (they could be formed in a fiber segment by processes of one or several cells, being displayed in a continuous or discontinuous manner) and their reticulate or fibrillar homogeneous structure.

Expansions of oligodendrocytes suffer shrinkage that allow them to easily occur with more or less fibrillar character, and this same contraction may explain a good part of the reticular formations (Figures 6.5, 7.13, 7.14, and 7.18), where we see enlargements of the cytoreticulum, simple or combined with neat filamentous formations.

The proved protoplasm shrinkage makes it difficult to know when they correspond to real differentiations of longitudinal fibers serving as binding to peri- and endomyelinic rings and cones (Figures 7.19, 8.2, and 8.5) and when they represent pleats and wrinkles of the subtlest membrane, so both possibilities are admissible. There are cases in which the longitudinal trabeculae are true small cords that continue with the thick annular reinforcements.

The rings that at intervals surround the myelin lining (Figures 7.13A, 7.14B, 7.19, 8.1, and 8.2) are never equidistant and often occur in pairs or series. They are formed by thick bands of condensed protoplasm in which a skein of parallel fibrils is quite often seen. The annular formations are extraordinary variegated, there being complete rings, simple and compound, with thick linking trabeculae, and incomplete rings, originating from the coil turns of gliocyte expansions, which, if circular and very close, form wide sleeves.

Sometimes the condensation of protoplasm is not reduced to thin annular strips, but in some tubes are cylindrical formations of variable length in which fibrillar striation is barely discerned and elastic membranes appear (Figures 7.1D and E and 8.3C–F).

The infundibular arrangements (cones located at intervals, with a more-or-less acute vertex, each fiber segment preferably directed in one direction, although it is not uncommon for vertices or bases to be face-to-face) with completely irregular distribution and made up of membranous septa stretching from the myelin surface to the axis cylinder, at times reflected over the same to ensheathe a certain amount of space (Figures 7.14B, 7.18D, 7.19I, 8.4A and C, and 8.5).

In the case of diaphragms formed by protoplasmic lamellae, they have neat fibrillar structure that makes them appear as nests or small baskets.

According to the preceding, it follows that in the central nerve tubes are formations absolutely identical to the rings, funnels, and reticular apparatuses characteristic so far of peripheral nerve fibers and that Cajal described better than anyone. And so that nothing is lacking, there are even identical arrangements to Nageotte's double bracelet, represented here by the completion and closing around the axon, with little radiating branches, of the sheaths corresponding to segments of the tube, formed by both oligodendrocytes.

Although Paladino and Montesano had a definite concept of the existence of neuroglial apparatuses regarding myelin, there are huge differences between what these authors described and reality. One first mistake is to consider astrocytes capable of forming sheaths attached to nerve fibers. Astrocytes insinuate, certainly, their appendages among the bundles of fibers, intertwining with them and curving more or less on their surface, but they never surround the bundles in the manner described by the cited authors. All formations closely attached to myelin belong to oligodendroglia.

Another error, and undoubtedly that of most importance, is to acknowledge the existence of cellular elements within the myelin sheaths, yet (a unique fact) without participating in the formation of its skeleton. We cannot know which kind of gliocytes correspond to those interpreted in this manner by Paladino and Montesano since the illustrations of their papers are extremely confusing. However, concerning small cells with scant cytoplasm and close enough to myelin sheaths that they could be thought to be within them, it is not unreasonable to suppose that they correspond to oligodendrocytes, whose processes Paladino and Montesano were unable to perceive even in outline. The latter author, however, saw in all probability some element of oligodendroglia in direct relation to nerve fibers.

# 7 Abnormal Changes in the Oligodendroglia

Having pointed out the possibility that the characters of peritubular neuroglial arrangements are modified by the effect of reagents, we must specify what we consider real and artificial regarding the morphological aspects that oligodendroglia show in normal and pathological circumstances.

## Autolysis

The high vulnerability of oligodendroglia, whose protoplasm is profoundly modified postmortem, is an easily observed fact. The autolysis phenomena observed by Cajal and described using the appropriate term of clasmatodendrosis, are not limited to protoplasmic and fibrous astrocytes but extend to every glia. Oligodendrocytes, therefore, often show morphological appearances, corresponding without any doubt to a more or less gradual process of cadaveric autolysis.

Human material, which is generally quite appropriate for studying oligodendroglia, when not belonging to early autopsied individuals shows the oligodendroglial corpuscles (silver carbonate staining) with more or less bulky soma, smooth contour or with small protrusions corresponding to its expansions, and their protoplasm appear swollen, clear, amorphous trabeculae being seen in it, which give them reticular appearance. The protoplasmic appendages, whose binding to the soma is not noticeable in all the elements, appear thickened, with ampullar bulkiness and very often fragmented. In the cerebral gray matter, in these conditions, it can be seen that the hydropic satellite corpuscles stand out in great relief and show, if they form pairs or groups, fusion of protoplasms in a multinucleate mass. Around them there are vestiges of their processes, but these are mainly recognized in the clear, smooth, or moniliform nerve fibers whose stained contour corresponds to the protoplasmic sheath formed by oligodendrocytes. The appearances hardly differ in the white matter, the cell bodies showing less relief and there being throughout the tissue very abundant clumps, more or less stained, corresponding to the fragmentation of the expansions of astrocytes and oligodendrocytes.

Faced with alterations of this kind, cannot easily be known whether they occurred during the disease or at the moment of agony; or if they are expression of cadaveric autolysis or reagent-dependent artifacts. Thus, to decide, to consider certain cellular aspects as real lesions, it is required great caution, and this will never be too much on assigning any of those to the pathological *substratum* of such-and-such a disease.

**Río-Hortega's Third Contribution to the Morphological Knowledge and Functional Interpretation of the Oligodendroglia.**
**DOI: http://dx.doi.org/10.1016/B978-0-12-411617-7.00007-9**
© 2013 Elsevier Inc. All rights reserved.

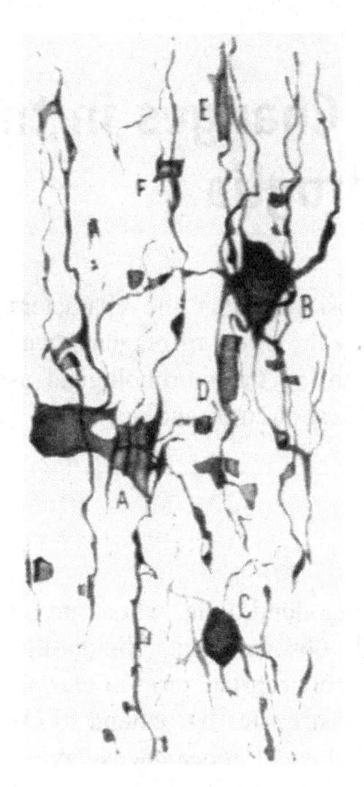


**Figure 7.1** Oligodendrocytes of the second type (cerebellar white matter of cat): A, enlarged process with signs of impression of the nerve tubes; B, cell expansions with plaques, rings, and tubular thickenings (D and F); C, cell element of the first type; F, ring formation.

However, in the supposedly normal human brain, instead of showing transparent spongy protoplasm, oligodendrocytes can appear intensely stained, with no sign of structure or with clotty and cloudy aspect. Believing that the normal texture corresponds to a loose protoplasm with almost imperceptible and finely granulated reticulum, which is the corresponding image for toluidine or silver carbonate good stainings, the transparent or cloudy swelling states should be interpreted as real lesions. But according to this criterion, we could hardly get to know the participation of oligodendroglia in nervous histopathology, because in almost every case disturbances such as those noted are found in greater or lesser degree.

## Progressive Phenomena

In principle, there is no doubt, the oligodendroglia intervenes in the pathological *substratum* of various nervous diseases with specific reaction forms since given it is a neuroglial variety it should not remain completely insensitive to the influences determining the amplifying strength in number and volume, or on the contrary, degeneration of the protoplasmic and fibrous astrocytes. However, in most processes determining the production of intense gliosis, it has not always been categorically proven that the amount of oligodendroglia is increased.

**Figure 7.2** Elements of the second type (A and B) involved in the formation of expansional plexus (C) (Microphot).

It is very frequent to see an apparent increase in the number of oligodendrocytes in brain regions whose nerve elements suffer atrophy and degeneration, on which the classical concept of neuronophagy was founded, that is, Marinesco's work. However, once the existence of nuclear pleiads (mainly oligodendrocytes) has been evidenced around normal nerve cells, hyperplasia of pseudoneuronophagic satellites could be argued.

After studying oligodendroglial behavior in many encephalopathies, we are convinced their responses (hypertrophy and hyperplasia) and degenerations follow the same behavior as astroglia—that is, *astrogliosis* is accompanied by *oligodendrogliosis* to a greater or lesser degree. The difficulty lies in discerning the early stages and even the intensity it reaches, because there has not yet been any tectonic study of oligodendroglia or a survey of their elements in different regions, which must necessarily be the starting point for all quantitative research.

In some cases of general paralysis in our experience (Figures 8.6 and 8.7), oligodendrogliosis is certain, since in any region of the cerebrum there is a greater number of cells than in the homologous of the normal cerebrum. The satellite pleiads of atrophic and degenerated neurons (Figure 8.6B and C) and of vessels (Figure 8.6A) occur in large amounts, offering morular aspects for the number of globular elements. This oligodendrocytic hyperplasia is accompanied not only by the typical astrogliosis of paralytic dementia but also by intense microgliosis (Figures 8.6D and 8.7B)—that is, by the two phenomena that should occur together and perhaps precede oligodendrocyte hyperplasia.

It seems unnecessary to say that if oligodendrogliosis accompanies cortical lesions, it should manifest itself later in the white matter. One cannot imagine to

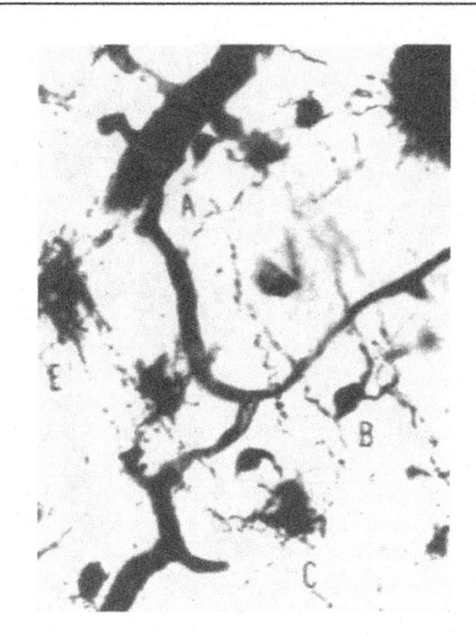


**Figure 7.3** Cerebral oligodendrocytes of the second type: A and B, with spiral expansions; C, tuberous element; E, element with thick cross-linked expansions (Microphot).

what extent the interfascicular oligodendroglia are capable of proliferating since the vast amount of elements filling the nerve fiber interstices makes discovering their increase impossible.

As to how the numerical increase of oligodendrocytes is produced, everything seems to indicate it occurs via direct division phenomena. While we have never observed mitosis, we have, however, observed several times (general paralysis, encephalitis of various origins), strangled (Figure 8.8A) and dumbbell-shaped (Figure 8.8C) nuclei as well as elements with two and up to three or four nuclei (Figure 8.8D and E). Nevertheless, when interpreting these forms with two or more nuclei one must remember that swollen transparent oligodendrocytes forming pleiads seem to constitute plasmodial masses which on a close examination reveal the contours of each protoplasm (Figure 8.8B). For this reason, when asserting the existence of corpuscles with more than one nucleus, we refer to cases where all of them were located on the same plane and only one cell contour was visible (Figure 7.4).

According to the aforementioned, we acknowledge oligodendroglia may proliferate moderately under adverse circumstances; yet to be able to assert the existence of *oligodendrogliosis*, sufficient intensity must be reached unless one can carry out numerical comparison on the normal homologous region. If this is not possible, one should stand by the existence of *astrogliosis and microgliosis*, taking into account that the most favorable destructive lesions for neuroglial hyperplasia provoke active microglial involvement at the same time, with the formation of rod-like cells and even granulo-adipose bodies. The example of general paralysis is very significant in this respect.

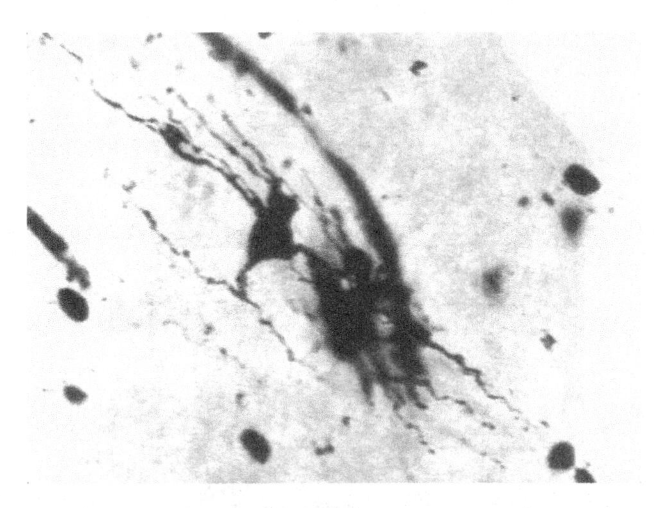

**Figure 7.4** Oligodendrocyte of the second type (cat cerebellum) with very long expansions of parallel course (Microphot).

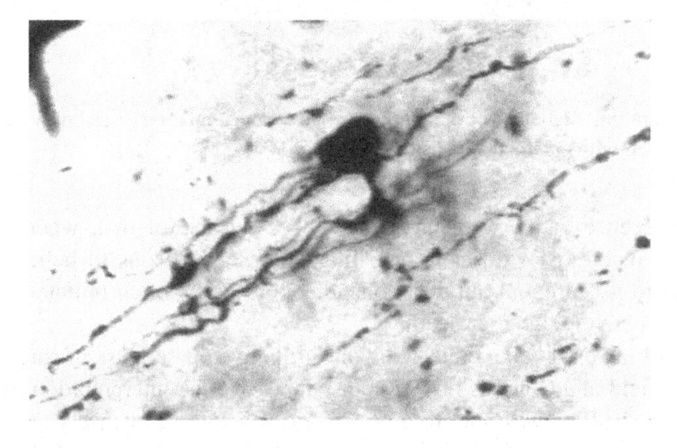

**Figure 7.5** Oligodendrocyte of the second type with satellite processes on nerve fibers (white matter of cerebellum) (Microphot).

## Degenerative Phenomena

We give such importance to the verification of microglial changes, as to the neuroglia (whether productive or degenerative), that in many cases their absence may constitute a useful argument for interpreting in a degenerative or autolytic sense, the swelling phenomena that occur in oligodendroglia, and which Penfield and Cone named *acute swelling*.

If microglia did not participate quickly in the destructive processes of the central nervous system causing neuroglial disturbances, we would not give importance

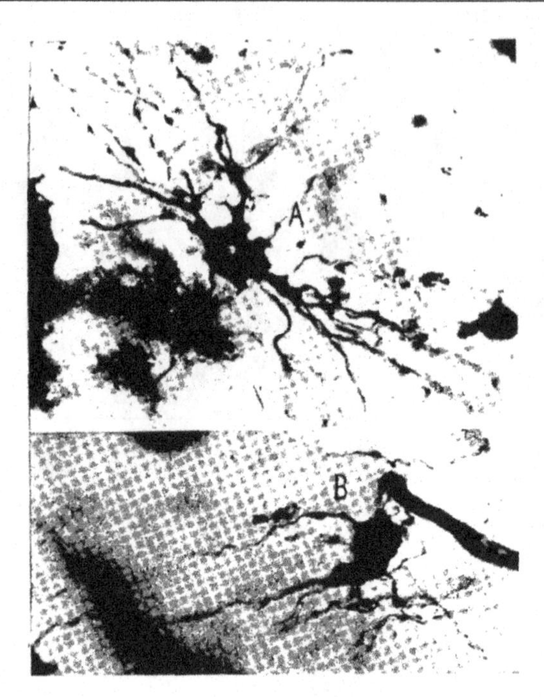


**Figure 7.6** Cerebral oligodendrocytes of the second type: A, tuberous element with tuftes of processes; B, cell with enlarged expansions (Microphot).

to their quiescence and macrophagic inactivity. Because of that, when microglial changes do not coexist with acute swelling, there are reasons to believe that this corresponds to a neuroglial autolysis phenomenon that is often initiated during the agony, to increase rapidly after death.

The fact of material being collected while still alive making use of surgical operations by Penfield and Cone, is not without interest for the interpretation of this process, as acute swelling was not present, whereas it appeared intensely in those cases where organic intoxication appeared slowly. We have also made observations on surgically obtained human material; however, the more or less pronounced phenomenon of oligodendrocyte swelling occurred thus (Figure 9.1), not significatively differing from that found in material from autopsies performed early. We had seen the swollen oligodendrocytes[1] in this latter, prior to Penfield and Cone's description, although it did not cross our minds to give them pathological interpretation, since the same thing or very similar occurred, almost without exception, in the human brain, without any relation whatsoever to the fatal illness.

In the circumstances pointed out by Penfield and Cone, oligodendrocyte autolysis began after the somatic swelling, which thus has a real acute character, as it happens at the end of life, and even specific, as to the type of hydropic swelling, whose pathological significance has yet to be determined.

[1] See Figure 1 of our note published in 1922, "¿Son homologables la glía de escasas radiaciones y la célula de Schwann?"

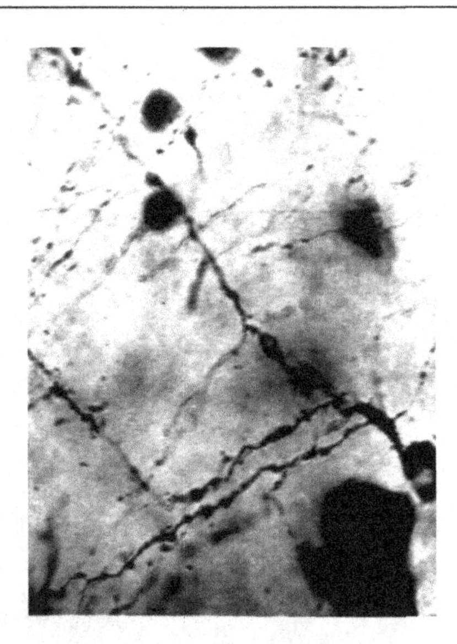

**Figure 7.7** Cerebellar oligodendrocyte of the second type with very long dichotomized processes (Microphot).

It can be deduced from the foregoing, that somatic and expansional swelling may occur in oligodendroglia by means of chemical–physical changes, favoured by the delicate protoplasm structure; however, it is almost always, simply agonic or cadaveric autolysis, with clasmatodendrosis, which does not exclude the possibility that in some cases, it may correspond to true pathological state, as yet insufficiently studied. For our part, after having worked on the participation of oligodendroglia in the *substratum* of various diseases and almost always finding endolytic phenomena, with somatic swelling and clasmatodendrosis, we could not say whether in general paralysis (Figures 8.6 and 8.7), senile dementia, epidemic encephalitis (Figure 8.11), tuberculous meningoencephalitis (Figure 8.12), syphilis, epilepsy, chorea, Friedrich's disease, tumors, hemorrhage, softening, hydrocephalus, surgical wounds, and so on, there are more intense swelling phenomena than in nonnervous diseases.

In material from laboratory animals, it would seem to be easier to make observations following the criterion of well-controlled equivalent images, but oligodendroglia do not usually show as visible swelling as in humans. Furthermore *Cajal's* clasmatodendrosis has a precociousness and intensity unknown in mammals, in human material. In these, in fact, there are generally no vast fields scattered with neuroglia clots largely of oligodendrocytic origin as exist in the human cerebrum.[2]

In short, it remains with contradictory judgements everything related to Penfield and Cone's acute swelling, which we would rather call *endolysis* or *transparent*

[2]While proofreading this paper, we have under study two cases of oligodendroglioma, one cerebral (central gyri) and another cerebellar, in which the swelling of extraneoplasic oligodendrocytes reaches great intensity. In the hemisphere white matter, very abundant elements with transparent swelling phenomena and vesicular or reticulate appearance are visible, and many (subependymal zone) have plasmolysis and are surrounded by irregular granules. All the white matter appears scattered with clear clots with irregularly annular shape, reminiscent of peritubular plaques and rings.

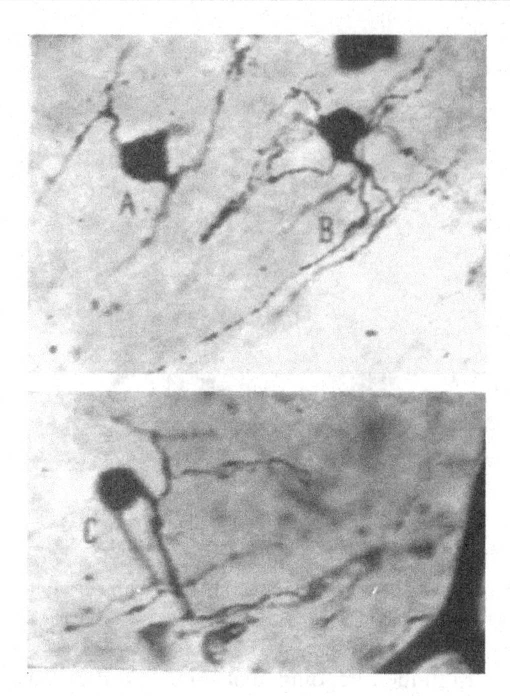


**Figure 7.8** Cerebral oligodendrocytes of the second type with appendages divided in T, accompanying nerve fibers: A, tripolar cell; B, multipolar cell; C, bipolar cell (Microphots).

*swelling* to better distinguishing it from another oligodendroglia aspect which could be named *cloudy swelling*.

Apart from the protoplasm cloudy appearance usually shown by oligodendrocytes in mammals, with silver carbonate, oligodendrocytes may suffer special pathological changes, that is, as a result of nervous disease. In some cases of tuberculous (Figures 8.9 and 8.10) and epidemic (Figure 8.11) encephalitis is where the mentioned alteration occurs more clearly, but it is not likely to constitute a type of injury characteristic only of inflammatory processes.

In their cloudy swelling state, the oligodendrocytes have a more or less bulky appearance, with visible relief, yet without the vesicular appearance characteristic of endolysis or transparent swelling. The protoplasm, whose thickened appendages are partly visible, appear diffusely stained, sometimes homogeneous and others finely granular, showing in some cases marginal or perinuclear vacuolization (Figure 8.11). The oligodendrocyte physiognomy so modified (Figure 8.10) is comparable to that of the ameboid elements, from which they differ mainly by the nuclear characters and size. Nuclei retain in general their normal size and structure, showing pyknosis in some cells.

The mentioned changes are accompanied by clasmatodendrosis (Figures 8.10 and 8.11), sometimes very intense, resulting in the presence of very copious dark clots scattered about in the network. But what seems more interesting in the cloudy swelling state is the fusion of protoplasms belonging to several elements that form irregular masses, with vacuoles, in which the cellular boundaries are erased

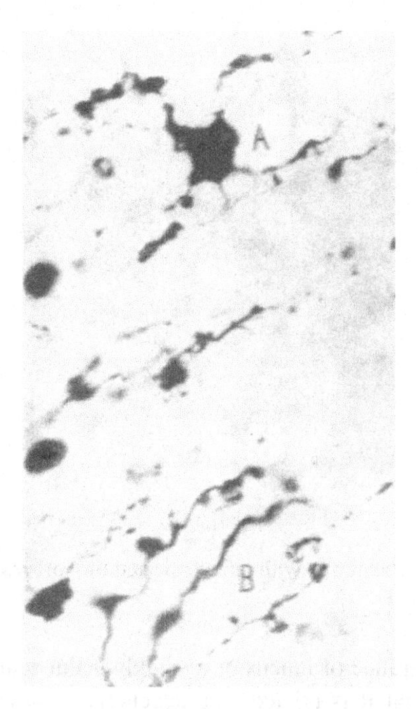


**Figure 7.9** Annular formations of the processes in oligodendrocytes of the second type: A, tuberous element with several extensions provided with widenings; B, cellular appendages forming rings (Microphot).

(Figure 8.10C). If fully confirmed, this cell fusion would be result of a plasmolysis phenomenon worthy of further study.

Oligodendrocytes sometimes differ from Alzheimer's amebocytes in showing the protoplasm turning into small clots (Figure 8.9). These clotty or granular types correspond to a variety of cortical ameboids described by Achúcarro and Gayarre in paralytic dementia and which we have found again in this disease and in tuberculous encephalitis.

In the horse trypanosomiasis (bad hip) is visible a special granular aspect of oligodendroglia (Figure 8.13) characterized by swelling of their elements, which appear filled with fine or thick granules and evidences of cell division (binucleated elements and isogenic pairs (Figure 8.13C, B).

The alterations mentioned herein do not normally lead to full cell destruction since nuclei remain intact or suffer only moderate pyknosis because of which the possibility of regeneration can be considered. Oligodendrocytes offer new protoplasmic aspects, but it cannot be known what relationship they have with the conditions of fixation and staining of the tissue.

There have been cases described, however, in which there seems to exist obviously a destructive trend in oligodendrocytes. We refer to studies initiated by Grynfeltt and Pélissier in senile processes on *mucoid degeneration*, which Bailey and Schaltenbrand have extended to Schilder's disease. For our part, we have observed some phenomena characteristic of mucoid transformation in a case of diffuse periaxial encephalitis and in others of hydrocephalus.

**Figure 7.10** A, Oligodendrocyte (A) with interconnected ring processes; B, apparently isolated rings (Microphot).

Regarding the appearance of mucus or a weakly stained substance, Grynfeltt and his confirmers admit that it is related to a degenerative process of the interfascicular oligodendroglia, whose elements are filled with mucoid matter. Our study on Schilder's disease has let us see in the injured areas typical rounded, swollen oligodendrocytes filled with a homogeneous substance stained faintly by toluidine (pink color) and refractory to mucus reagents. In the case of hydrocephalus (Figure 8.14), the areas close to ventricular lining showed many elements in a more or less graded swelling state (A–E), containing a substance dimly stainable by toluidine blue and silver carbonate in yellowish color. Oligodendrocytes, converted into little bladders, show a more or less visible contour, but their boundaries are not erased in any case, lacking, therefore, any evidence of protoplasmic fusion into multinucleated masses as seen by the above-mentioned authors.

These details are insufficient to think of mucoid degeneration, since it is known that metachromatic reactions in several elements may occur in the nervous tissue. Myelin, which in some circumstances obtains with toluidine blue an identical staining to that of mucus, could lend itself to confusion if the hypothesis were accepted that it could fuse and infiltrate the closest interstitial elements, or that myelin being a product elaborated by oligodendrocytes, is retained in them accidentally, swelling the protoplasm. When observations are not sufficient to judge with solid ground, every hypothesis is acceptable and rejectable. Because of this, considering ourselves *de visu* uninformed about the process of the alleged mucoid degeneration, we omit any definitive opinion.

In diffuse periaxial encephalitis, the mucous change of oligodendrocytes is not the most important factor but their greatly reduced number parallel myelin disappearance. This should not be viewed merely as a consequence, since if, as abundant evidence shows, oligodendroglia effectively participates in the production and

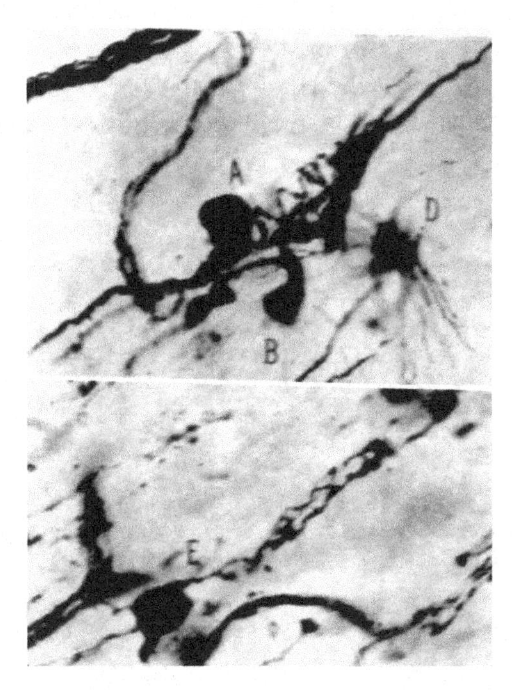


**Figure 7.11** Oligodendrocytes of the third type of cerebral white matter of cat: A, element with a spiral expansion surrounding a nerve tube; B, pyriform element; D, fibrous astrocyte; E, cell process with a reticular formation around a nerve fiber (Microphots).

maintenance of the myelin while the medullary sheath is surrounded and partitioned by the oligodendrocyte protoplasm, when the latter is changed the former must be altered, from which a pathogenic relationship of obvious interest may be inferred. But oligodendroglia being homologous of Schwann cells, the behavior of both classes of perimyelinic elements in pathological processes injuring nerve fibers must be identical. In the case of Schilder's disease, the hypothetical possibility must be considered that the absence of myelin is not caused by degenerative phenomena but by agenesis of oligodendrocytes. Insufficient or unevenly distributed, these could not meet the needs of nerve fibers, which would develop poorly and disappear over time.

The cytopathological outline that we are making, in which we purposefully omit all indications made by the authors without firm knowledge of microglia and oligodendroglia in their respective forms and functions, would be incomplete if we did not speak about certain activities attributed to oligodendrocytes, such as the retention of fat and iron and the phagocytic faculty.

## Phagocytic Intervention

The possibility of forming granulo-adipose bodies at the expense of oligodendroglia has been discussed by several authors, whose approaches have been recalled in

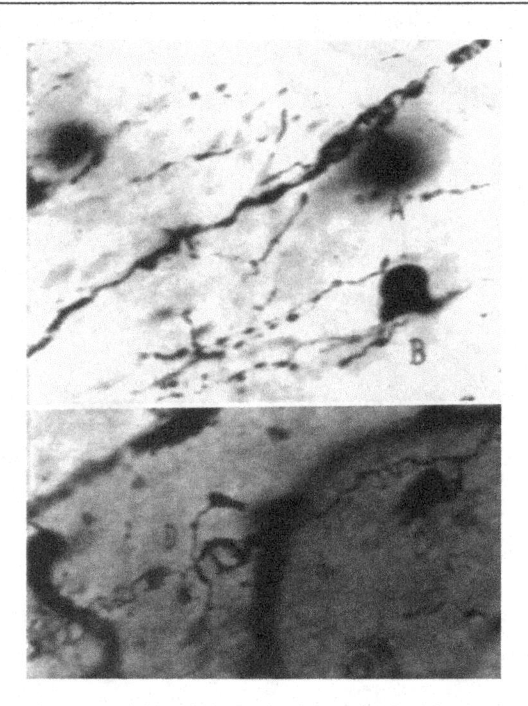


**Figure 7.12** Various elements of the third type in the cerebral white matter of cat: A, monopolar oligodendrocyte with expansion that accompanies and ensheathes a nerve fiber; B, cell of the first type; C, pyriform oligodendrocyte with process (D) that forms a delicate reticulum around a medullated tube (Microphots).

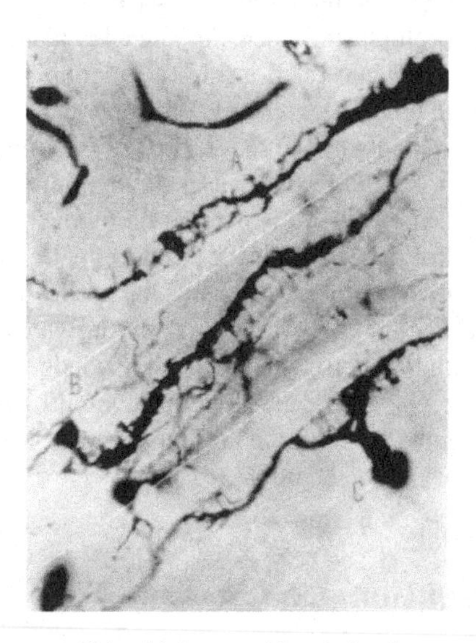


**Figure 7.13** Oligodendrocytes of the third type with spirals, rings, and reticula around thick medullated tubes of the cerebral peduncles: A, retiform process with wide rings; B, monopolar cell with a wide extension provided with ansiform endings; C, pyriform cell with spiral branches (Microphot).

chapter 2. The similarity between small nuclei of oligodendroglia and those rounded of microglia makes it possible to mix them up when one has no exact knowledge of the volume, shape, and chromatic richness of the different nuclei. However, things do not happen like that. Despite the somewhat insecure opinions of various authors and that of Meduna (too hastily published, since more experience, based on better stainings, would make him change his approach), microglia are in charge of a phagocytic role in which oligodendroglia are not involved at all. Whichever morphological approach one may wish for both cell species (which are not even genetically related), whether the microglia were considered as genuine neuroglia forming part of a syncytium, a work of imagination, and the oligodendroglia as an element without processes, that is, as apolar glia, about what there is no doubt is that the microglia work as macrophage, charge with fatty products and form granular bodies, whereas the oligodendroglia, on the contrary, are not a phagocyte and eventually enclose small lipid droplets endogenous or exogenous (this is debatable) without ever acquiring the character of granulo-adipose bodies.

Struwe and other authors have shown the presence of fat droplets in oligodendrocytes, and their observations are not debatable, since, in fact, under various circumstances lipoids appear in the vicinity of some nuclei, either as isolated droplets or as lateral small piles. The problem lies in determining the genesis of such fatty products, given the rejection of the hypothesis that it corresponds to an active phagocytic phenomenon, it may be thought to result from metabolism disruptions (similarly to how accumulations of pigment are formed in astroglia) or else true cell degeneration.

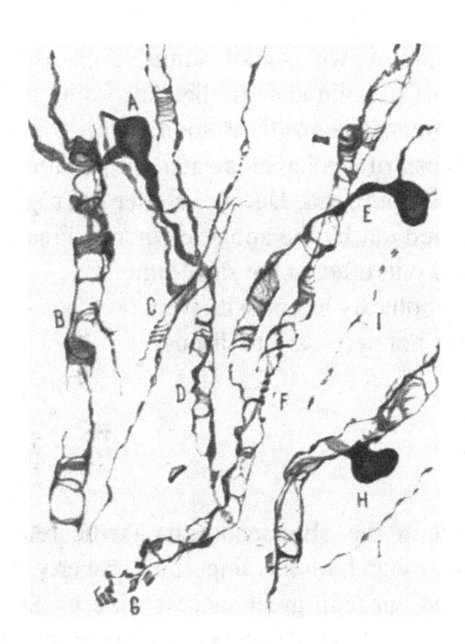


**Figure 7.14** Oligodendrocytes of the third type of cerebral white matter: A, cell with sheaths around three myelinated tubes; B, rings and infundibula; C, spiral fiber; D and F, retiform aspects; E, process divided into an acute angle around two tubes; G, little rings for thin fibers; H, monopolar cell with an expansion divided in T that ensheathes a nerve fiber; I, appendages with plaques and little rings on very thin fibers.

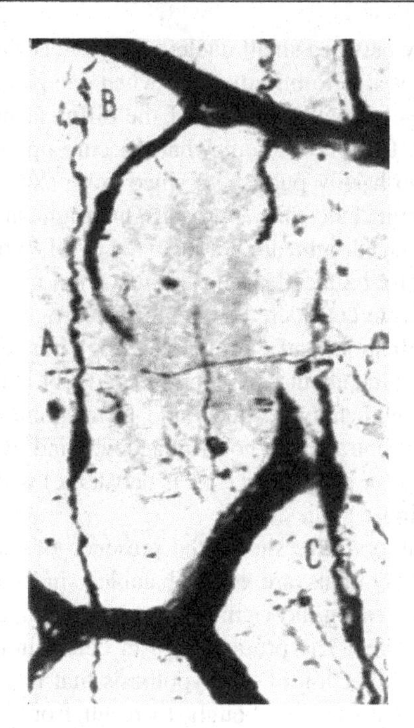


**Figure 7.15** Oligodendrocytes of the fourth type in spinal cord white matter of cat: A, bipolar cell; B, ring formations; C, cell with enlarged retiform extensions (Microphot).

Against Meduna's opinion, we cannot admit, even as remote possibility, the phagocytic participation of the oligodendroglia. According to Meduna, as a result of nerve cell alteration the microglia swell; as soon as this microglial swelling reaches a certain point the process of apolar element multiplication begins, which brings about phagocytosis of the microglia. During or after microgliophagy starts the neuronophagy, likewise carried out by the apolar elements. Finally, microgliophagy and neuronophagy are carried out, often at the same time.

While this original hypothesis in opposition to the ideas supported by a legion of researchers, we think it is not necessary to discuss it.

## Sideropexy

Another activity studied in the oligodendroglia is the retention of cerebral iron (*Eisenspeicherung*). The research on this important property in normal and pathological state has been carried out with great success: first by Spatz and Metz and then by Struwe, Malamud, and Wilson. According to Spatz and Metz, oligodendroglia normally have the capability to retain iron, and in pathological cases this property would be shown more intensely. Our experience about this participation in storage of

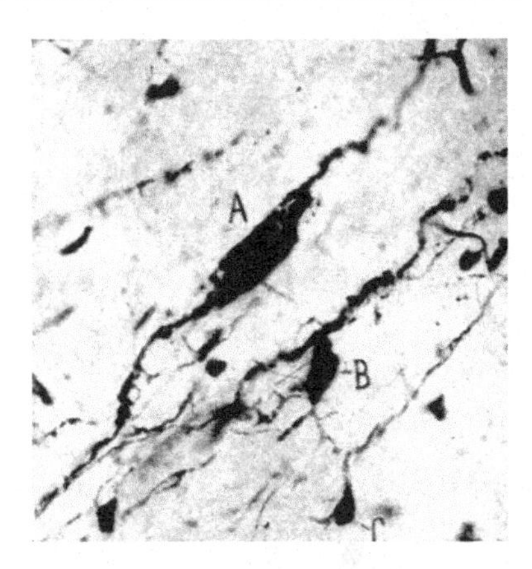


**Figure 7.16** Oligodendrocytes of the medulla of the cat: A, cell of the fourth type with flattened soma on a medullated fiber; B, cell of the third type with expansion in T; C, second-type cell (Microphot).

normal iron, stuck in the cerebrum or from hematic disintegration (hemorrhages), is quite limited but enough for us to be able to judge it.

According to our observations (Figure 9.1), oligodendroglia are diffusely stained very pale blue by Perl's and Turnbull reactions, but the staining is limited to the surface, because of a very fine granular deposit on the cell membrane. When the iron reaction is made repeatedly in the same sections (following our technique for microglia and other elements of the reticuloendothelial system), in some cases we see a subtle precipitate of iron particles in the trabeculae of the spongy protoplasm and only exceptionally is the odd small true blue granule seen.

In pathological cases, the amount of iron contained in the oligodendroglia does not seem to be increased. Thus, in general paralysis (*Paralyseeisen*), we have not been able to see more intensely stained elements in whole, or with more granules, whereas microglia offered their soma and almost all their processes perfectly dyed in intense blue, showing besides iron clots included in the protoplasm.[3]

In the vicinity of hemorrhagic foci, where microglia hoard nearly every product of hematic disintegration (*Blutzerfalleisen*), giving the most intense colorations of hemosiderin, oligodendrocytes and astrocytes frequently present a more or less intense positive Perl's reaction, acquiring a diffuse blue dye in which some small clots of iron stand out.

After studying the circumstances in which iron occurs in neuroglial elements, we think that microglia are the only cell species that collects and stores iron from hematic origin by specific function, properly macrophagic, and that oligodendrocytes

[3]The fixative (simple formalin or bromide) strongly influences the ferric reactions of microglia and oligodendroglia.

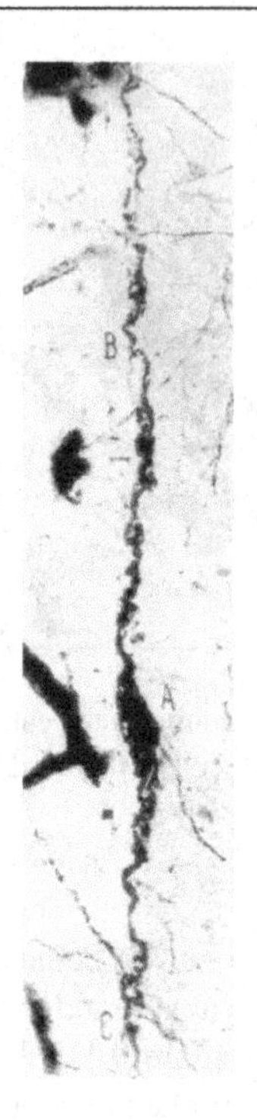


**Figure 7.17** Oligodendrocyte of the fourth type, spinal cord of cat: A, flattened soma on a myelinated tube; B and C, reticular, annular and infundibular formations (Microphot).

and astrocytes do not collect ferric substances in an active manner, since to do so they would require ameboid movements that are incompatible with the syncytial idea (German school) or plexus (Spanish school) of their expansions. As the products from hemoglobin in the proximity of the hemorrhagic foci spread out they partly reach oligodendrocytes and astrocytes, passively allowing their protoplasm to become impregnated if hemosiderin is in sol state, while retains it superficially if in gel state. Therefore, is a simple adsorption phenomenon.

The preceding brief analysis on the morphological changes of oligodendroglia has no other purpose than to note the guidelines followed for their study, noting the difficulty of obtaining definitive results. Agony or cadaveric changes, deformations of reactive origin, and the unreliability of the study techniques hinders the clarification of the cytopathological truth. For this reason, transparent or cloudy swelling, granular and ameboid degeneration, the abnormal fixation of fat and iron are many other contentious issues that sometimes took too early resolution. Researchers have hastened to discover abnormalities without prior accurate knowledge of the oligodendroglia in normal state, and we have likewise followed the same path, and whilst we do not feel the need to rectify some opinions, in others we fully submit ourselves to that resulting from the best verification of facts.

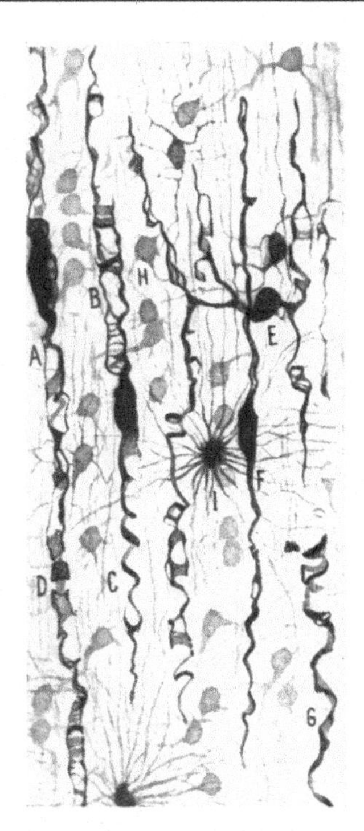


**Figure 7.18** Oligodendrocytes of the spinal cord white matter of cat: A, B, and F, elements of the fourth type with reticula and rings (B), spirals (C and G) and funnels (D); E, third-type cell; H, cell of first type; I, astrocyte.

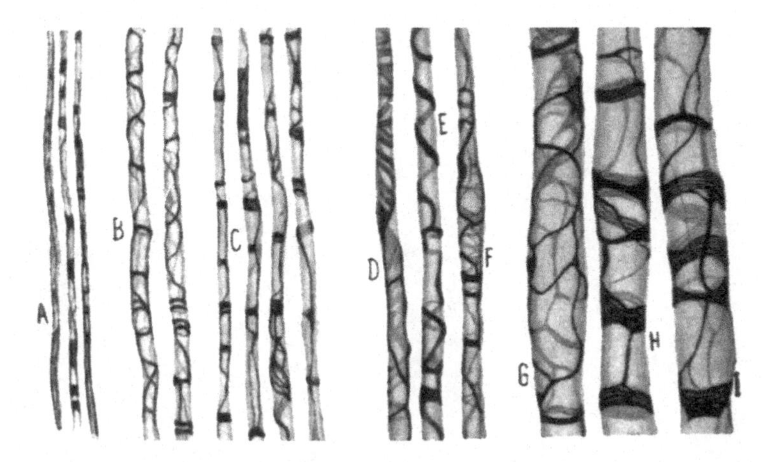


**Figure 7.19** Different aspects of the perimyelinic sheaths formed by oligodendrocytes: A, thin fibers; B–F, fibers of medium thickness; G–I, thick fibers; A, subtle sheaths with little rings and plaques; B, fine reticula with turns of coil and rings; C, rings with variable thickness connected by flanges; D, laminar arrangement; E, spiral arrangement; F, reticular arrangement; G, loose reticulum between two rings; H, dense and laminar rings with fibrillar structure joined by trabeculae; I, membranous funnel.

# 8 Morphological and Genetic Kinship Between Astrocytes and Oligodendrocytes

In this work, polarized in a morphological sense, we could not, even if we wanted, fully address the genetic study of oligodendroglia, since the techniques previously used to demonstrate embryonic neuroglial elements gave null results with respect to oligodendrocytes, and those we are now running have not yet been applied methodically to embryos of newborn animals.

It is not necessary, however, to know fully the genealogy to discern the lineage, as what is not achieved with observation is obtained by logical deduction.

When the existence of corpuscles without expansions, which could represent a kind of germinal glia or undifferentiated elements from mesodermal origin (since their ambiguous characters prevented classification) was believed it was not possible to assign them to the neuroglia without violating the classical concept of these. Today, when a new technique discovered a rich plexus of expansions in those apparently apolar elements, one cannot think they constitute an undifferentiated glia or mesoglia. Oligodendrocytes belong to a species with many varieties of the glia genus which is recognized by its generic and specific characters.

The preceding pages include a detailed description of somatic and expansional forms, which show the fundamental difference separating oligodendroglia from

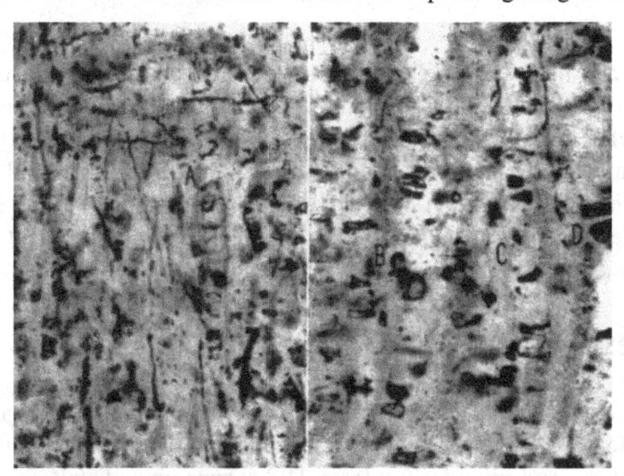


**Figure 8.1** Sets of ring formations of the human medulla (two different magnifications): A, B, C, series of rings corresponding to three nerve tubes; D, widened rings with fibrillar structure (Microphots).

Río-Hortega's Third Contribution to the Morphological Knowledge and Functional Interpretation of the Oligodendroglia.
DOI: http://dx.doi.org/10.1016/B978-0-12-411617-7.00008-0
© 2013 Elsevier Inc. All rights reserved.

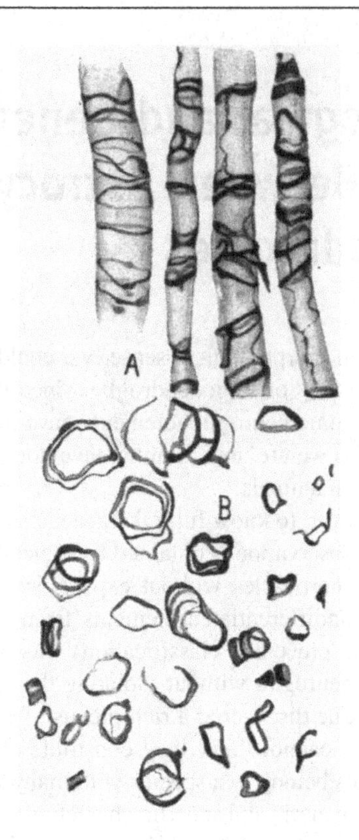


**Figure 8.2** Ring formations of myelinated tubes of the human medulla: A, fiber sights along; B, cross section of a fascicle.

protoplasmic and fibrous astroglia. The attempt to fill this separation space between each element with other morphologically equidistant cells would be in vain, which clearly indicates that oligodendroglia are a neuroglial type with high and peculiar differentiation. There is not thus a true morphological transition, and this demonstrates two important facts: (1) oligodendroglia come from the same stem but not from the same branch as astroglia and (2) from their origin, they adapt morphologically to the function they must perform.

## Histogenesis

Everything seems to indicate that in the early *neuroepithelium*, two genera of elements are differentiated: *neurocytoblasts* and *glioblasts*. On detaching, the latter give rise to *astroblasts and oligodendroblasts*. The former immediately acquire connections with mesodermal elements (vessels and sheaths), and the latter are placed in intimate relationship with nerve fibers.

The emergence of both sister species of neuroglia is performed simultaneously or in successive periods, accompanying the development of the nervous elements and their neurites and of the vascular system.

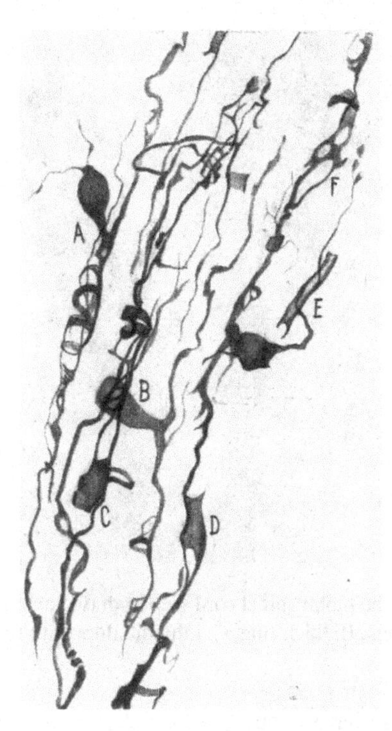


**Figure 8.3** Oligodendrocytes of the third type of the cerebellar peduncle: A, B, glomerular and retiform arrangements; C, D, E, laminar arrangements; F, fenestrated arrangements.

The *astroblasts* (intimately connected to vessels via direct somatic insertions—perivascular glia—or long pedicles) evolve in the protoplasmic (*astrospongioblast*) and fibrous (*astroinoblast*) direction. The *oligodendroblasts* associated with nerve fibers retain their delicate protoplasm and form special tonic apparatuses.

The protoplasmic neuroglia (astrospongioglia) adapt themselves to the presumed trophic or secretory function (which does not exclude that of insulation of the nervous elements). The fibrous neuroglia (astroinoglia) differentiate for the supporting function (which does not exclude the secretory one), and the oligodendroglia are arranged around nerve fibers to insulate them and serve as a trophic vehicle by elaborating myelogenous matter.

The genealogy of neuroglia can be summarized as follows:

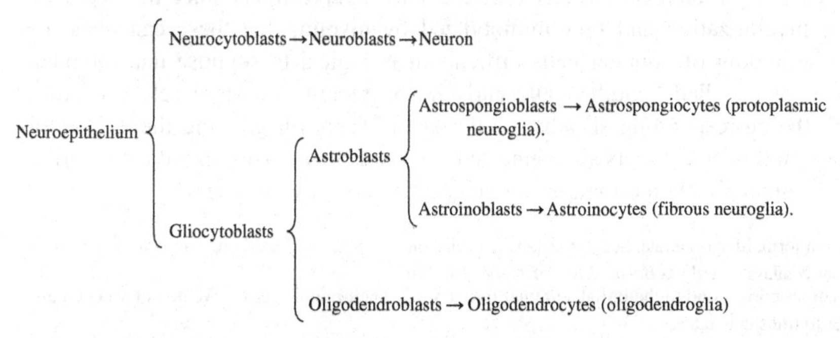

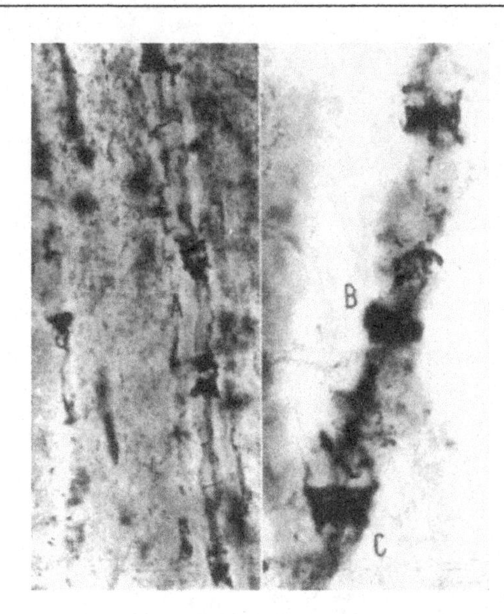


**Figure 8.4** Nerve tubes in the rabbit spinal cord seen with two magnifications: A, tube with fibrillar infundibula and rings; B, thick ring; C, infundibulum with fibrillar structure (silver carbonate*) (Microphot).

The preceding names[1] have a purely didactic purpose, and we do not seek to impose them. Corresponding to elements of uninterrupted development with imperceptible transitions from the original epithelial type to the definitive forms, whatever nomenclature is adopted has to be rather rough and capricious. Although we have said that we do not know the early developmental stages of oligodendrocytes, the preceding genealogy being work half observational and half logical deduction, the solidly grounded theoretical concept will remain secure against every possible discussion.

The oligodendroglia seem to be the neuroglial species whose development finishes later since in newborn and young mammals (Figure 9.2) one can still witness their formation in large numbers. As they are designed to combine with nerve fibers, their largest increase corresponds to the development of same.

However, considering stereotropic cell movements are not easy in branched corpuscles; that the migration of oligodendroblasts in very long series, sliding among the nerve bundles, is only possible if the number of processes is minimal; that there are very abundant elements solidly connected with nerve fibers since the beginning of their myelinization and thus immobilized by myelin; that these can make the interstitial sliding of younger cells difficult; it is logical to suppose that oligodendrocytes are installed hypothetically early in the vicinity of nerve cells and fibers to form the corresponding sheaths for the latter. Then, on growing the axis cylinder, staggered new elements are connected and once these have formed their fibrillar sleeves, remain almost motionless and dragged by the growing fibers.

*Fixation in formol-ferric sulfate. See, for technical details, our 1925 paper, Condrioma y granulaciones específicas de las células neuróglicas *Bol. de la Soc. Esp. de Hist. Nat.*

[1] These names correspond to the initial and final forms of each neuroglial species. We do not accept names that refer to transient states.

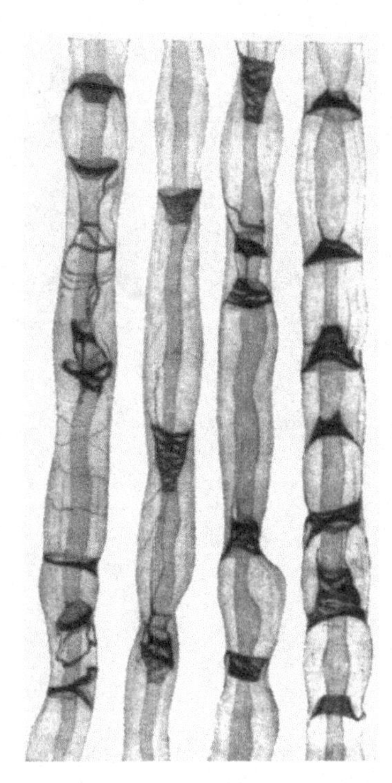

**Figure 8.5** Annular and conical formations around the nerve tubes in the spinal cord of rabbit (silver carbonate stain).

On arising, oligodendroblasts would migrate into some space, acquiring relationships with one or many nerve fibers, allow themselves to be carried out by these in their growth, and leave the starting point, where new elements would be willing to act in the same manner. The long series of oligodendrocytes would not, therefore, be formed by lined-up corpuscles (Figure 9.3) moving clumsily to reach their destination. The shape of oligodendrocytes would depend in part on the number of original appendages and on those arising in subsequent periods, and in part on the growth of nerve fibers, which would stretch the protoplasmic expansions more or less, and on their thickening, which would widen the mentioned expansions and provoke the development of annular and reticular tonic apparatuses.

The oligodendrocytes acquire peculiar differentiations early, during incipient developmental stages, when they only have generic characters, offering similarities to the astrocytes, making difficult to distinguish between them in some cases; nevertheless, as they complete their development and specific features emerge, the similarity to protoplasmic and fibrous neuroglia disappears.

## Transitional Forms

To impose the finding of transitional forms among different elements as a requirement for acceptance of their genetic kinship, is an error, acknowledged by many

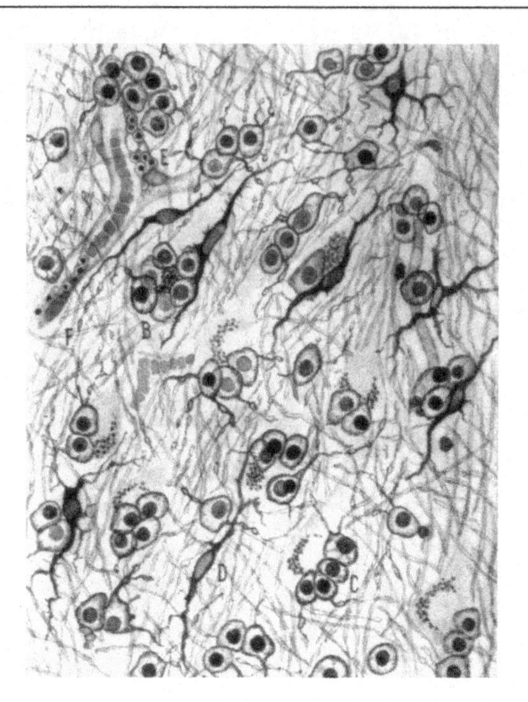

**Figure 8.6** Transparent swelling and proliferation of oligodendrocytes in the cerebral cortex in dementia paralytica: A, vascular pleiad; B, C, perineuronal groups (pseudoneuronophagy); D, microglia; E, endothelial cell inclusions; F, nerve fibers covered by oligodendrocytes' processes (silver carbonate stain).

scholars. Indeed, one can reach the greatest absurdities by finding such forms of transition. Thus, for example, no one will be able to prove that microglia and neurons derive from a common branch or transform one in the other, and yet many nerve corpuscles could be misinterpreted as microglia.[2] Nor is it demonstrable that oligodendrocytes have something in common with some short-axon nerve elements and with the grains of the cerebellum, despite sometimes striking morphological coincidences among them. When study is done by means of the Golgi method or any of its variants, an oligodendrocyte located in the cerebellar cortical layer only differs from the grains in the manner of how the latter finish their brief expansions. If any short-axis-cylinder cell were deprived of its axon, then it could only be distinguished from some of the oligodendrocytes with difficulty. The protoplasmic form thus does more to obscure than to elucidate the problem.

Checking shape transitions can be useful provided no absolute values were given to the positive results and even less to the negative ones.

For example, it has been said in order to deny the mesodermal lineage of microglia that there are no transitions between them and lymphocytes; however, this

[2]We refer the reader to the masterful work of Cajal "Textura del sistema nervioso" and in it the figures 106, 172, 179, and 254 (volume 2), featuring images of nerve cells very similar to microglia, and figures 332 and 339, showing aspects of short-axis-cylinder elements very similar to oligodendroglia.

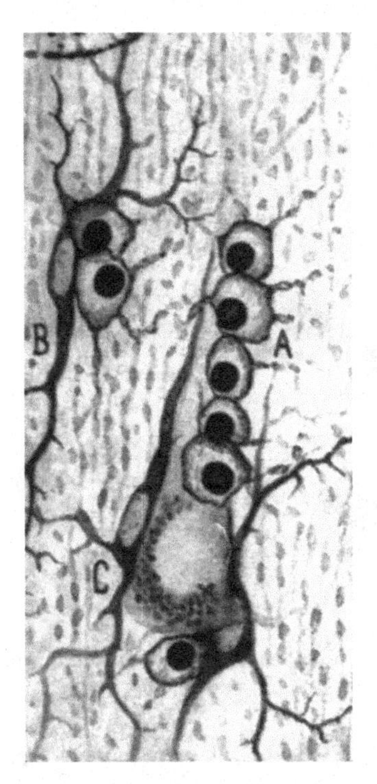


**Figure 8.7** Proliferation, swelling, and clasmatodendrosis of neuronal satellite oligodendrocytes (A) in general paralysis; B, C, microglia (silver carbonate stain).

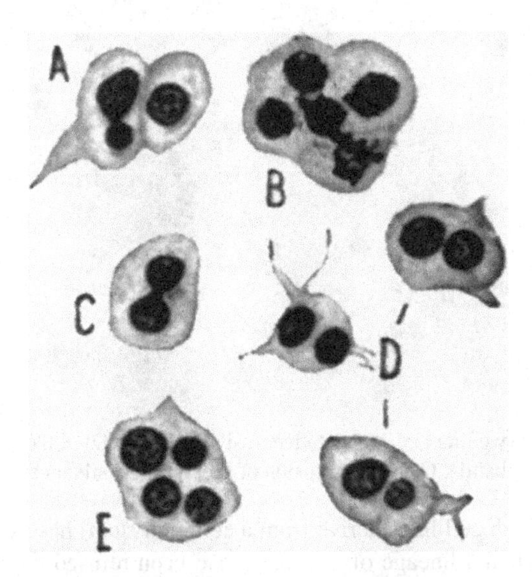


**Figure 8.8** Cortical oligodendrocytes of the epidemic encephalitis: A, C, nuclei with amitotic division; B, morular arrangement; D, binucleate cells; E, tetranucleate cell.

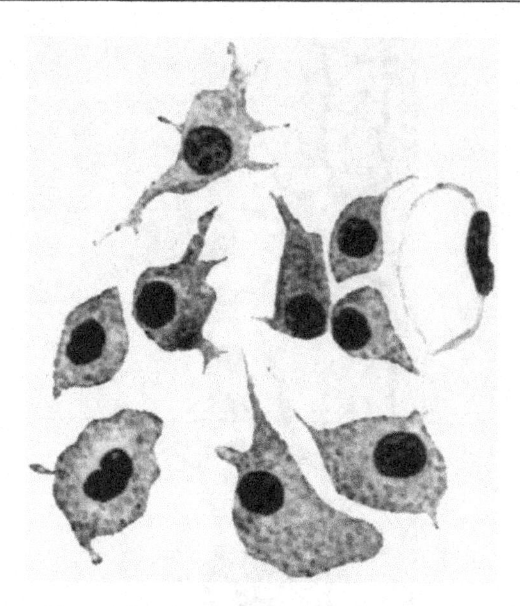

**Figure 8.9** Cloudy swelling of oligodendrocytes in tuberculous meningoencephalitis (silver carbonate stain).

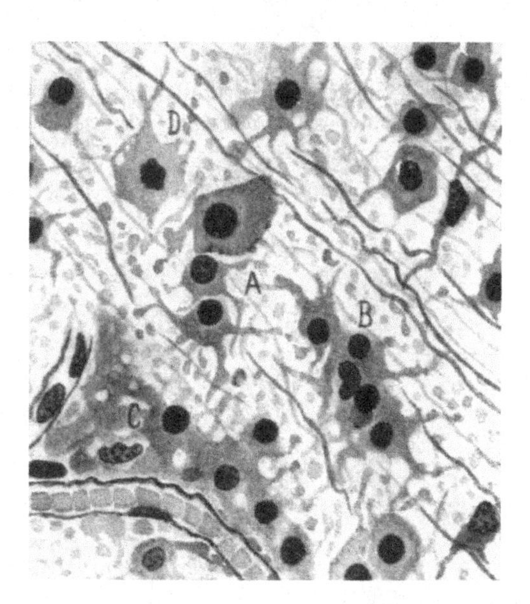


**Figure 8.10** Cloudy swelling and clasmatodendrosis (A, B, and D) of the oligodendroglia in a case of meningoencephalitis; C, apparent fusion of protoplasms (silver carbonate stain).

(which is true as both cell lines derive from a common stem) has as much conviction as if the neuroepithelial lineage of astrocytes had been refused because there are no transitions between them and neurons.

The structure also has a relative importance for cell systematization. Regardless of the trophic organization of protoplasm, which differs only in relation to cell size and nucleus location, and paying attention to the latter, there is no doubt that oligodendrocytes have round and chromatin-rich nuclei that strongly resemble large lymphocytes; appearing process-free like these, they lend themselves to confusion. But nor can it be doubted that, among such nuclei of oligodendrocytes and cerebellar grains (as between these and lymphocytes), apparent differences are almost null. Because of this, the nuclear characters, not being too precise, are not used to determine the genetic kinship of various cellular elements.

The same could be said about the dyeing affinities. Not only the stainability of heterogeneous elements but also the unstainability of related elements with a particular technique have led and will lead to large errors. Even in the most favorable case when a specific technique for a genus of cells is used, one can say all those stained belong to it, yet not denying there are nonstained cell species of the same genus. Such is Cajal's auric method which exclusively reveals neuroglia, yet not all species of neuroglia, leaving oligodendroglia to one side of the staining capacity of the method.

There are no real transitional forms between astrocytes and oligodendrocytes, since neither the former (Figure 9.4B and D) acquire connections with nerve fibers or the latter with the vessels (Figure 9.4A), but there does exist a cell type with ambiguous characters that we are describing briefly.

It mainly settles in the cerebral white matter (Figures 9.5B and C and 9.6A–C), is interspersed with astrocytes and oligodendrocytes, and shares with the latter the size and shape of the soma, but it differs from them by the number and character of its expansions. These are very numerous (Figures 9.7B and C, 9.8B and C, and 9.9A)

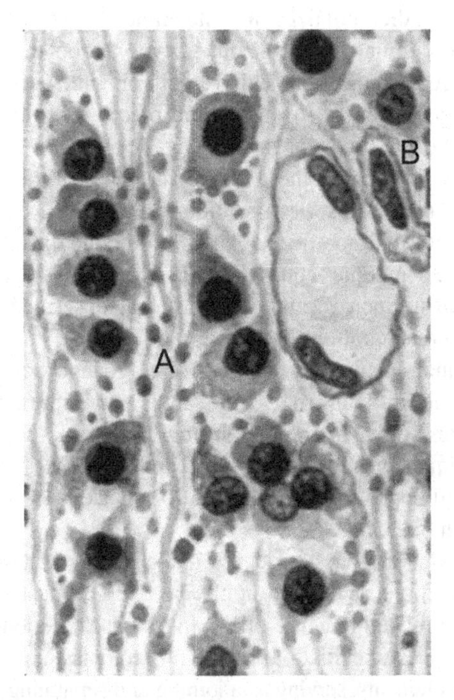


**Figure 8.11** Swelling and clasmatodendrosis of oligodendroglia (A) of the optic thalamus in epidemic encephalitis; B, granulo-adipose body (silver carbonate stain).

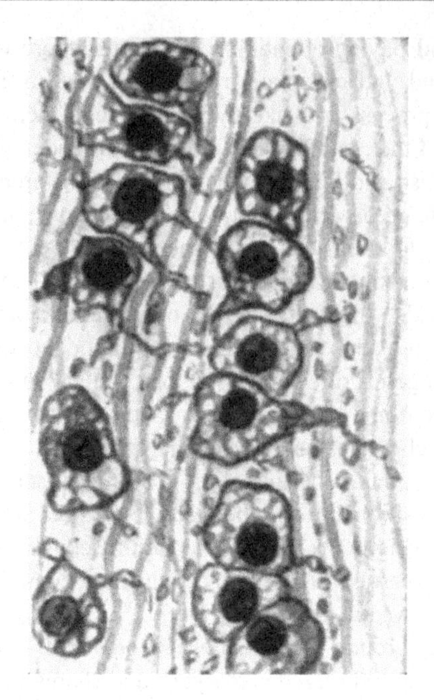

**Figure 8.12** Reticular and vacuolar swelling of the interfascicular oligodendroglia in a case of tuberculous meningoencephalitis (silver carbonate stain).

and appear not very long, dichotomized at acute angles several times and with a semiprotoplasmic character similar to that of mixed-type astrocytes. The set of appendages from where a direct connection with vessels or with nerve fibers is not seen (which cannot deny its existence), forms a kind of crown, whose diameter is very much smaller than that of protoplasmic astrocytes.

The most interesting thing about these dwarf astrocytes of the white matter, whose significance we fail to interpret, is the possibility of mistaking them for oligodendrocytes in the staining with the modified Golgi method, since by the somatic size and number of processes are sometimes equal (Figures 9.7D, C, and B and 9.8B and C).

We give little value to such forms, and with that described we seek to satisfy those who show more consideration for them.

Knowing the morphological characteristics of oligodendrocytes, their neuroglial category is not arguable, although if it were, their neuroepithelial origin could be confirmed in the cat and other mammals, during embryonic development and after birth, using our techniques or others more elective. Nevertheless, researchers show a unique tendency toward opposition when trying to replace ideas which were fashionable with more modern ones.

In the case of oligodendroglia, considered apolar, some authors express belief in the mesodermal origin more or less explicitly, perhaps because of ignorance of the processes that are close related to nerve fibers. Indeed, the criterion for identifying neuroglia was based on the classic definition: cells with protoplasmic or fibrous expansions interspersed with nerve elements, serving as a joining and separating medium for them at the same time. It seemed the oligodendroglia did not meet these conditions until now.

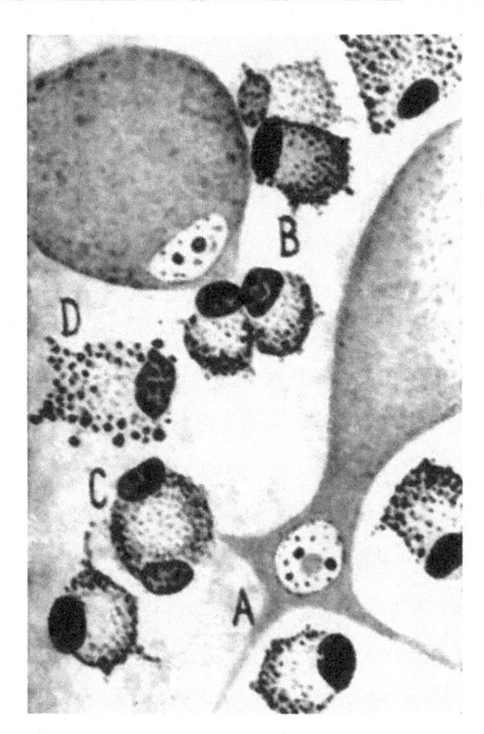


**Figure 8.13** Granular swelling of the oligodendroglia in the horse trypanosomiasis (bad hip): A, nerve cell enormously swollen; B, isogenic pairs of oligodendrocytes; C, binucleated corpuscle; D, elements in plasmolysis (silver carbonate stain).

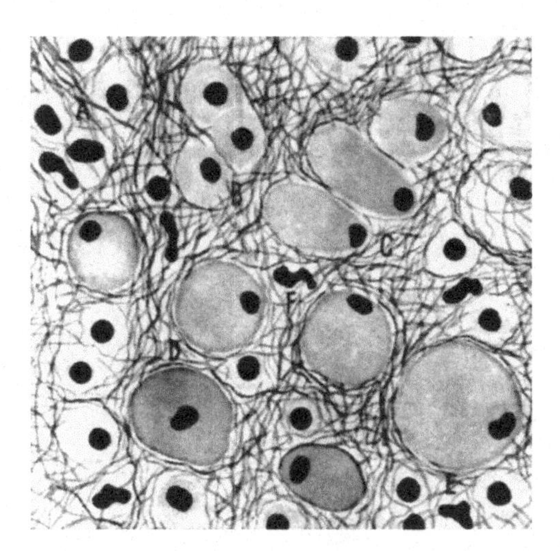

**Figure 8.14** Oligodendrocytes of the subependymal zone of the cerebrum in a case of hydrocephalus: A, B, C, D, E, different degrees of mucoid degeneration; F, microglia (silver carbonate stain).

# 9 Morphological and Physiological Parallelism of the Oligodendrocytes and Schwann Cells

In Chapter 6, we provided a detailed description of the morphological varieties of oligodendrocytes, demonstrating that they constantly acquire close relationships with nerve fibers. These relationships are manifested by the formation of periaxonic sleeves that reach very diverse development, being extremely simple in the thin myelinated fibers (with respect to unmyelinated ones, we have no data to assert or deny the existence of neuroglial cases) and acquiring progressive complication in the thick medullated tubes.

In the white matter of the cerebral and cerebellar hemispheres, oligodendrocytes are located along the nerve fascicles in whose interstices they form long series. In general, cell bodies remain at a distance from fibers ensheathed by their long expansions; although they may show traces of imprint, they are only rarely flattened on the medullated tubes.

In the thick nerve bundles of the cerebral and cerebellar peduncles, pons, medulla and spinal cord, and nerve roots, although there are many elements with rounded soma and very short thick expansions, for many other immediate fibers, bipolar cells closely attached to the myelin sheaths also abound. So oligodendrocytes sometimes send expansions to distant nerve fibers, other times to very close fibers and finally others, flatten themselves on the fibers to which they are destined. The morphological resemblance to Schwann cells, which is very small in the first and second types, gradually increases (third type) and then becomes complete (fourth type).

The careful study of the morphological peculiarities of each variety of oligodendrocytes and of the progressive organization of the perimyelinic protoplasmic sheaths reveals, with absolute evidence, that between Schwann cells and the elements morphologically most distant from them is an uninterrupted gradation in which the progressive organization of myelin sheaths is followed step by step, starting from the thinnest fibers of the cerebral cortex and reaching to the thick tubes of the medulla and spinal cord.

One might think that Schwann cells are oligodendrocytes in the highest degree of differentiation in which the symbiotic organization with nerve fibers is completed. This organization appears staggered along nerve pathways, and if it were possible to follow a medullated fiber from its origin in the cerebral cortex and through the pyramidal tract up to the nerve, it would be seen associated with a chain of oligodendrocytes, which at a certain time of their organization could not be differentiated from Schwann cells.

**Río-Hortega's Third Contribution to the Morphological Knowledge and Functional Interpretation of the Oligodendroglia.**
**DOI: http://dx.doi.org/10.1016/B978-0-12-411617-7.00009-2**
© 2013 Elsevier Inc. All rights reserved.

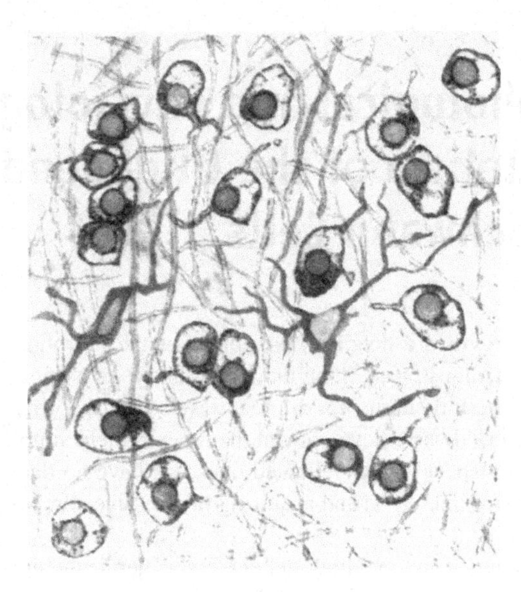

**Figure 9.1** Oligodendroglia and microglia in the vicinity of a cerebral injury (Prussian blue reaction).

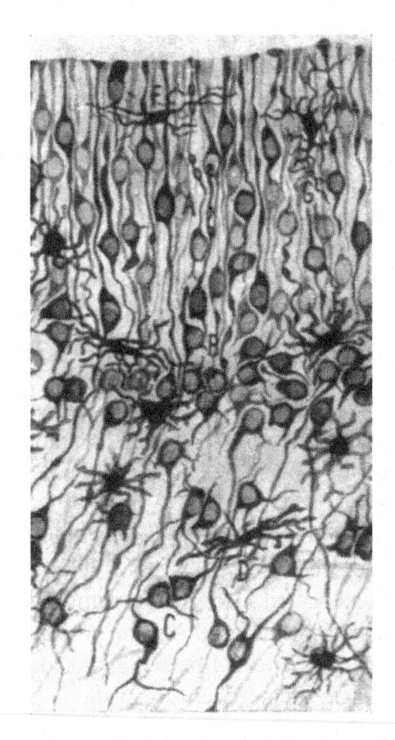

**Figure 9.2** Ependymal zone of a cerebral lateral ventricle of a newborn kitten: A, gliocytoblasts; B, oligodendroblasts; C, oligodendrocytes; D, E, F, G, microglial elements (silver carbonate stain).

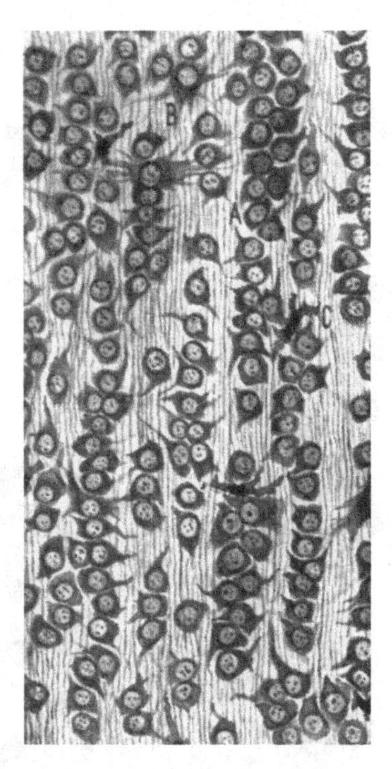


**Figure 9.3** Serial arrangement of the oligodendroglia in the spinal cord of a newborn kitten; A, oligodendrocytes; B, astrocyte; C, microgliocyte (silver carbonate stain).

The most recent studies of these cells, to which Cajal and his disciples have contributed to a superlative degree, unveiled in them a set of details. Although discussed with regard to their relations with the nuclei, it cannot be doubted that they correspond to peri- and endotubular protoplasmic differentiations, forming sometimes extremely complex supporting apparatuses for myelin.

The formations characteristic of Schwann cells are identical to those found in the most differentiated oligodendrocytes. In both kinds of corpuscles occur fenestrated or reticular membrane-shaped sheaths with fibrous and annular reinforcements and septa or diaphragms sectioning the columns of myelin at intervals likely further extend around the axons as a subtle and intimate sheath.

If the images obtained by Cajal and Sánchez of the peripheral nerve tubes are compared (formol-uranium method), verified by many researchers, with those of the central nerve tubes supplied with the modified Golgi method and silver carbonate is found such an identity that sometimes it could not be said whether they belonged to oligodendrocytes or to Schwann cells.

Although much has been written on Schwann cells, the morphological interpretation of peri- and endomyelinic formations associated with them has not been achieved univocally, being possible that our findings regarding oligodendroglia organization could help to state it precisely.

Studies by Ranvier, Nemiloff, Nageotte, Cajal, Doinikow, Sánchez, Schroeder, and so on have proven in Schwann cells the following: (1) a perinuclear protoplasm

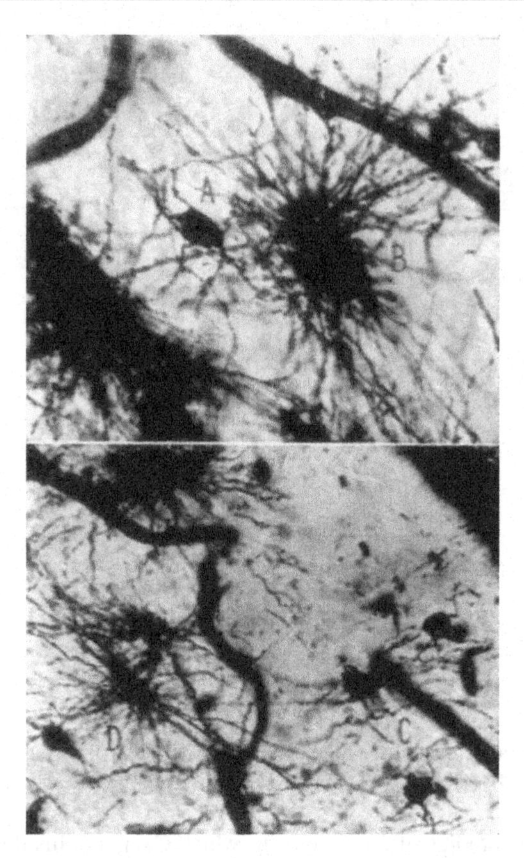


**Figure 9.4** Astrocytes (B, D) and oligodendrocytes (A, C) in the cerebral white matter of dog (Microphot).

that continues all around the myelin as broad longitudinal bands and transverse trabeculae (Cajal); (2) a system of differentiated filaments in relation to these trabeculae and, just like them, more or less branched (Nemiloff, Sánchez); (3) rings (Ségall) joined to the preceding fibers by means of longitudinal trabeculae in relation to the edge of Lantermann's incisures or independent of them (Cajal); and (4) infundibuliform apparatuses in relation to Lantermann's incisures formed by a spiral filament (Rezzonico, Golgi) or a plexus of anastomosed filaments, largely transverse (Nageotte, Cajal), joined by a cement.

However, the peculiarity of the four referred formations of Schwann cells offering different dye reactions has led amorphous protoplasm, reticular and annular formations, and infundibular apparatuses of Lantermann's incisures to be described separately. Our research on central nervous tubes demonstrates that both the sheaths and septa representing Lantermann's incisures in them are formed by oligodendrocyte protoplasm in a varying degree of differentiation and chromatic reactions, which cause that in some cases only rings or infundibula appear, and in others their unions with the oligodendrocyte expansional protoplasm are clearly visible.

**Figure 9.5** Different oligodendroglial varieties of the cerebral white matter of cat: A, oligodendrocytes; B, C, dwarf astrocytes; D, fibrous astrocyte (Microphot).

Nevertheless, as the same peri- and endomyelinic organization of central and peripheral nerve tubes (as any existing differences are of degree, not of kind), the affirmation reticula, fibers, rings and infundibula are differentiated parts of the Schwann cell is not arbitrary.

Regarding the function performed by oligodendrocytes, it would not be logical if we did not identify it with that characteristic of Schwann cells. The morphological equivalence corresponds to the same functional activity. What matters now is to find out what this is.

As the role of Schwann cells related to myelin has not yet been determined, as some think it is housed in the protoplasm of those in the form of inclusions, others claim myelogenesis is a function of the axis cylinder. Both hypotheses have arguments in favor, one of the most important for the defense of the axonal myelin origin from its degeneration after destruction of nerve fibers.

Sacristán (J. D.) has demonstrated in Schwann cells the presence of protoplasmic granulations with undoubted characteristics of secretory granules, and this fact could be important in the problem we are dealing with. Oligodendroglia also have a granular content, well developed, corresponding to astrocyte gliosomes; like these, the content accumulates in the somata and is distributed along the expansions.

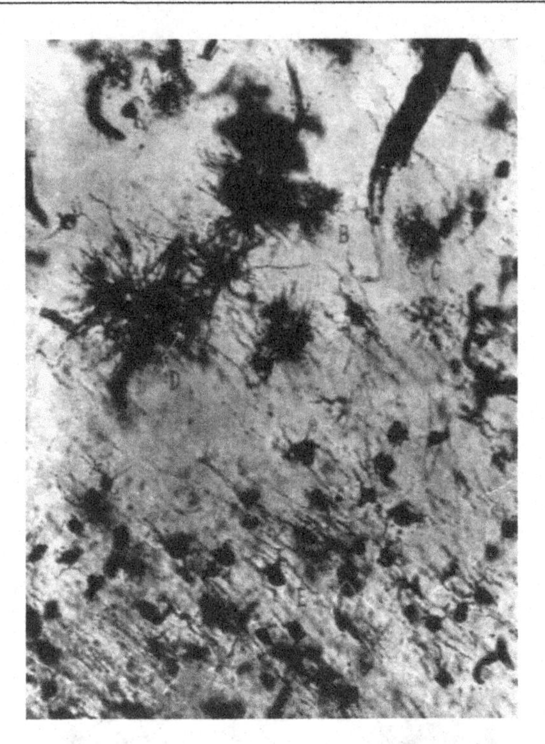

**Figure 9.6** Neuroglial cell varieties in the cerebral white matter of cat: A, B, C, dwarf astrocytes; D, fibrous astrocytes; E, oligodendrocytes (Microphot).

Such granulations are particularly abundant during animal development when myelinization of nerve centers is at its maximum activity and when growth in the length and thickness of nerve fibers is high.

In a newborn kitten (Figure 9.10), very abundant granulations of variable size are found along nerve fibers in the cerebrum, cerebellum, and spinal cord, arranged in irregular series on outward forms of appendages of oligodendrocytes that are poorly colored. At a certain moment, such granules lose visibility due to the myelin sheath already being formed and stained simultaneously. It would seem that myelin is formed by confluence of the aforementioned granulations, but a confirming demonstration of this very important fact has not yet been achieved in an indisputable manner.

Signs of secretory activity of oligodendroglia suggest that it is involved in myelogenesis, but such intervention could be direct or indirect. In the first case (which constitutes our working hypothesis) myelin would be an oligodendrocyte secretion, or, more precisely, a production (incretion) of their protoplasm, which will always retain relationships. In the second case, oligodendrocytes would provide to nerve fibers the necessary trophic materials for the production of myelin at their expense.

If myelin did not disintegrate as a result of nerve fiber degeneration, denoting a close association with them, perhaps in the sense of assuming myelin as a formation

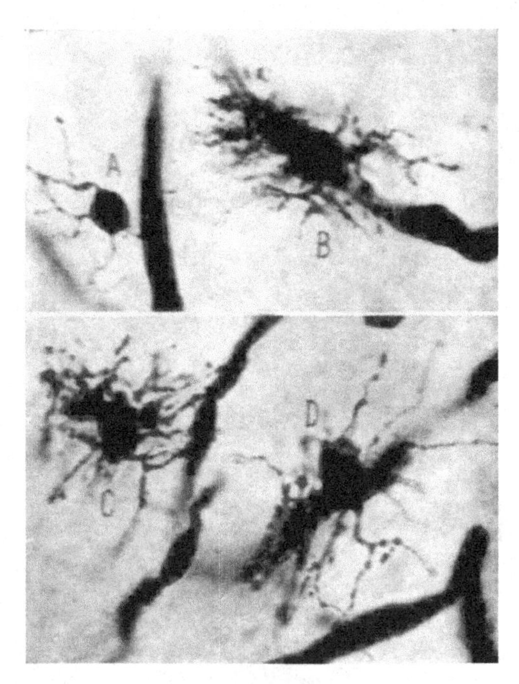


**Figure 9.7** Transitional forms between oligodendrocytes (A, D) and dwarf astrocytes (B, C) (Microphot).

of Schwann cells adjacent to the axons the problem would already have been solved. The degeneration phenomenon would then be explained assuming the nerve fibers and Schwann cells joined in symbiosis.

There is no data yet to confirm the behavior of oligodendrocytes in secondary degenerations, but what takes place in Schilder's periaxial encephalitis lets us suppose that they do not remain intact when nerve tubes disappear. The disappearance of medullated nerve fibers in the areas affected by periaxial encephalitis is linked to a huge reduction in the number and degenerative states of oligodendrocytes (mucoid degeneration). There is an evident relationship between both phenomena, but the circumstances of simultaneity or independence in which they occur cannot be known.

Moving away now from the rugged terrain of hypothesis, where it is easy to get lost, and limiting ourselves to objective reality, we can state: (1) that oligodendrocytes form myelin sheaths homologizable with Schwann sheaths and septa equivalent to Lantermann's incisures, and (2) that in these formations occur specific granules that are an expression a protoplasm production process, more intense during myelinization of nerve pathways.

The first fact leads us to consider oligodendrocytes as elements of protection (sheaths with fibrillar and annular reinforcements) and of support (intercalated septa) for myelin, which probably are at the same time vehicles for trophic substances for axons (Lantermann's incisures have been interpreted in this sense).

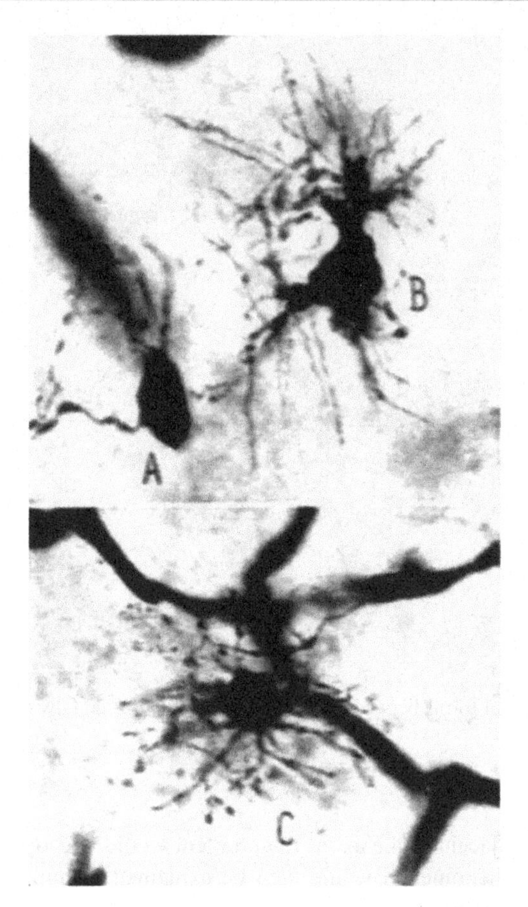


**Figure 9.8** Transitional forms (B, C) between oligodendrocytes and dwarf astrocytes: A, oligodendrocyte incompletely stained (Microphot).

The second fact leads us to think that the protoplasmic production is related to myelogenesis through the direct incretion of myelin (which seems the most likely to us) or by providing materials to build it.

In any case, oligodendrocytes and Schwann cells are in a very broad sense morphologically, physiologically, and, in all likelihood, histogenetically brothers.

## Conclusion

In the study, we have just carried out, there is undoubtedly a part referring to *what is seen* in oligodendrocytes (using the most selective techniques) and another questionable part concerning the morphological and physiological *interpretation* of their relationships with nerve fibers.

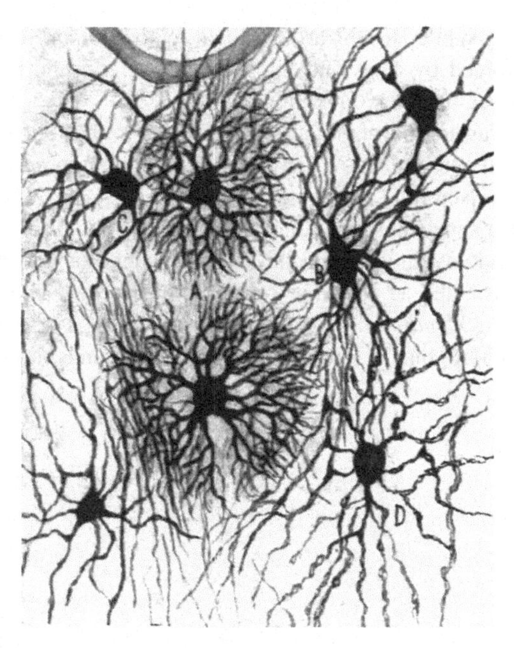


**Figure 9.9** Somatic analogy and expansional dissimilarity between dwarf astrocytes (A) and oligodendrocytes (B, C, D).*

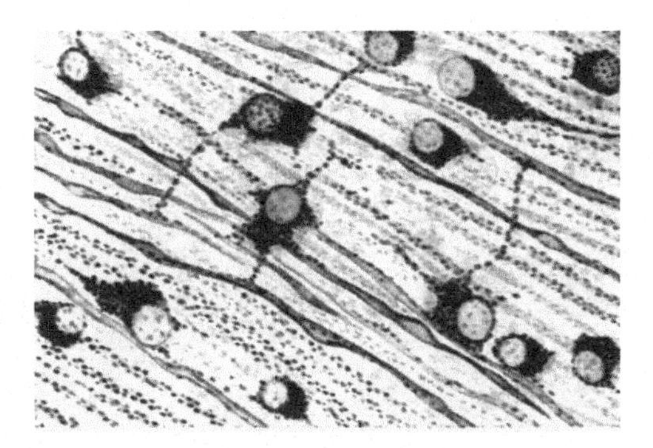

**Figure 9.10** Granular contents of oligodendroglia and granulations of nerve fibers in process of myelinization. Newborn cat cerebrum (silver carbonate stain).

We have reported numerous facts that powerfully enlighten an important sector (which until now remained in shadow) of the intimate texture of the central nervous system, and we have used the tectonic details to build what we regard as an extremely logical hypothesis regarding the functional value of oligodendroglia.

The preceding pages demonstrate that oligodendrocytes, whatever their size, shape, and number of protoplasmic expansions, establish close relationships with

*This and all figures in which the technique has not been indicated correspond to stainings with silver chromate.

nerve fibers forming subtle sheaths around them, with laminar, fenestrated, or reticulate aspects reinforced by rings and interannular flanges, and by infundibuliform intramyelinic diaphragms in the thick tubes.

All these structures are morphologically equivalent to those in Schwann cells and sheaths of peripheral nerve tubes. When they reach maximum development, they do not differ at all from Nemiloff's networks, Ségall's rings, and infundibula characteristics of Lantermann's incisures as studied by Rezzonico and Cajal.

The peri- and endomyelinic organization of oligodendrocytes allows them to be physiologically considered as *Schwann cells of the nervous centers.* Real Schwann cells are, in turn, *oligodendroglia of the nerves.* And if logic does not fail, we must assume that oligodendroglia and Schwann cells have the same blastodermic origin.

# References


Achúcarro N, Gayarre M: La corteza cerebral en la demencia paralítica con el nuevo método del oro sublimado de Cajal, *Trab. del Lab. de Inv. Biol.* XII:1–38, 1914.

Alberca R: Intervención precoz de la microglía en las heridas experimentales de la médula del conejo, *Bol. de la Soc. Esp. de Biol.* XI:81–88, 1926.

Alberca, R: Estudio histopatológico de la encefalitis experimental. *Tesis doctoral.* Madrid, 1928.

Bailey P: Further remarks concerning tumors of the glioma group, *Bull. Johns Hopkins Hospital* XL:334–389, 1927.

Bailey P, Cushing H: *A classification of the tumours of the glioma group on a histogenetic basis with a correlated study of prognosis*, Philadelphia, 1926, J.B. Lippincott.

Bailey P, Hiller G: The interstitial tissues of the central nervous system: a review, *J. Nerv. Ment. Dis.* LIX:337–361, 1924.

Bailey P, Schaltenbrand G: Die muköse Degeneration der oligodendroglia, *Deutsche Zeitschr. für Nervenheilk* XCVII:231–237, 1927.

Bergmann, RAM: Die Zellen von Hortega und ihre Färbung. *Dissertation. Utrecht*, 1927.

Cajal SR: El aparato endocelular de Golgi de la célula de Schwann y algunas observaciones sobre la estructura de los tubos nerviosos, *Trab. del Lab. de Inv. Biol.* X:221–246, 1912.

Cajal SR: Fórmula de fijación para la demostración fácil del aparato reticular de Golgi y apuntes sobre la disposición de dicho aparato en la retina, en los nervios y en algunos estados patológicos, *Trab. del Lab. de Inv. Biol.* X:209–220, 1912.

Cajal, SR: Estudios sobre la degeneración y regeneración del sistema nervioso, Vol. I, 1913.

Cajal SR: Contribución al conocimiento de la neuroglía del cerebro humano, *Trab. del Lab. de Inv. Biol.* XI:255–315, 1913.

Cajal SR: Algunas observaciones sobre la mesoglía de Robertson y Río-Hortega, *Trab. del Lab. de Inv. Biol.* XVIII:109–127, 1920.

Cajal SR: Beitrag zur Kenntnis der Neuroglia des Gross - und Kleinshirns bei der progressiven Paralyse mit einigen technischen Bemerkungen zur Silberimprägnation der pathologischen Nervengewebes, *Zeitschr. für die ges. Neurol. und Psychiatrie* C:738–793, 1926.

Cajal SR, Tello Muñoz JF: *Elementos de histología normal y de técnica micrográfica*, Madrid, 1928, Tipografía artística.

Catanni J: L'appareil de soutien de la myeline dans les fibres nerveuses peripheriques, *Arch. Ital. de Biol.* VII:345–356, 1886.

Cerletti H: Estudi recenti sull'istogenesi della nevroglia, *Annali dell Istit. psichiatrico della R. Univer. di Roma* IV:221–234, 1907–1908.

Collado C: Participación de la microglía en el substratum patológico de la rabia, *Bol. de la Soc. Esp. de Biol.* IX:175–191, 1919.

Creutzfeldt HG, Metz A: Die morphologische und funktionelle Differenzierung der Neuroglia, *Zentralblatt für die ges. Neurol. und Psychiatrie.* XXXVIII:416–418, 1924.

Creutzfeldt HG, Metz A: Über Gestalt und Tätigkeit der Hortegazellen bei pathologischen Vorgängen, *Zeitschr. für die ges. Neurol. und Psychiatrie.* CVI:18–53, 1926.

Da Fano C: Some recent methods for the study of neuroglia, *J. R. Microsc. Soc.* XLVI(2):89–102, 1926.

Doinikow B: Beiträge zur Histologie und Histopathologie des peripheren Nerven, *Histol. Histopathol. Arbeiten Großhirnrinde* IV:445–630, 1911.

Grynfeltt E: Mucocytes et leur signification dans les processus d'inflammation chronique des centres cérébrospinaux, *Comp. rend. des séanc. de la Société de Biologie.* LXXXIX:1264, 1923.

Jakob A: Über die feinere Histologie sekundären Faserdegeneration in der weissen Substanz des Rückenmarks (mit besonderer Berucksichtigung der Abbau vorgänge), *Histol Histopathol. Arbeiten Großhirnrinde* V:1–181, 1912.

Jakob, A: *Normale und pathologische Anatomie und Histologie des Großhirns*, Vol. I. Leipzig, 1927. F. Deuticke.

Jimenez Asúa F: Die Mikroglia (Hortegasche Zellen) und das reticuloendotheliale System, *Zeitschr. für die ges. Neurol. und Psychiatrie* CIX:354–379, 1927.

López Enríquez M: Oligodendroglía de las vías ópticas, *Bol. de la R. Soc. Esp. de Hist. Nat.* XXVI, 1926.

Marano A: I rapporti del nevroglio con le cellule e le fibre nervose, *Annali di Neurol.* XXIX:1–6, 1911.

Marchesani O: Die Morphologie der Glia im Nervus opticus und in der Retina, dargestellt nach dem neuesten Untersuchungsmethoden und Untersuchungsergebnissen, *Von Graefes Arch. für Opht.* CXVII:575–605, 1926.

Meduna L von: Beiträge zur Histopathologie der Mikroglie, *Arch. für Psychiatrie u. Nervenkr.* LXXXII(2), 1927.

Metz A: Die drei Gliazellarten und der Eisenstoffwechsel, *Zeitschr. für die ges. Neurol. und Psychiatrie* C:428–449, 1926.

Metz A, Spatz H: Die Hortegaschen Zellen (das sogenannte dritte element) und über ihre funktionelle Bedeutung, *Zeitshr. fiir die ges. Neurol. und Psychiatrie.* LXXXIX:138–170, 1924.

Montesano G: Circa il comportamento dello "Scheletro nevroglico di Paladino nelle fibre nervose delle diverse zone ed aree del midollo spinale, *Riv. sperim. di freniatria* XXXVIII(2-3), 1912.

Nageotte J: Incisures de Schmidt-Lantermann et protoplasme des cellules de Schwann, *Comp. rend. des séanc. de la Société de Biologie* XLVIII:39–42, 1910.

Nemiloff A: Einige Beobachtungen ueber den Bau der Nervengewebes bei Ganoiden, etc, *Arch. für mikrosk. Anat.* LXXII:575–606, 1908.

Nemiloff A: Über die Beziehung der sog. Zellen der "Schwannsche Scheiden" zum Myelin in der Nervenfasern von Säugetieren, *Arch. für mikr. Anat. und Entwickl.* LXXVI, 1910–11.

Paladino G: De la continuation de la névroglie dans le squelette myelinique des fibres nerveuses et de la constitution pluricellulaire du cylindraxe, *Arch. Ital. de Biol.* XIX:26–32, 1892.

Paladino G: Dei limiti precisi fra il nevroglio e gli elementi nervosi nel midollo spinale e di alcune questione istofisiologiche che si vi riferiscono, *Boll. della R. Acc. med. di Roma* XIX(1), 1893.

Paladino G: Ulteriori studi sui rapporti fra il nevroglio e le fibre e le cellule nervose nell'asse cerebro spinale dei vertebrati, *Rend. della R. Acc. delle Sc. fis. e mat. di Napoli*(8–12), 1900.

Paladino G: Ancora dei più intimi rapporti fra il nevroglio e le cellule e le fibre nervose, *Rend. della R. Acc. delle Sc. fis. e mat. di Napoli*, 1908.

Pélissier G: *Syndrome wilsonien consécutif a la névraxite épidérmique. Contribution a l'étude de la dégénérescence mucocytaire de la névroglie*, Montpellier, 1924. L'Abeille.

Penfield WG: Oligodendroglia and its relation to clasical neuroglia *Bol. de la R. Soc. Esp. de Hist. Nat.*, 1925, *Brain* LIII:430–452, 1924.

Penfield WG, Cone W: Acute swelling of oligodendroglia. A specific type of neuroglia changes, *Arch. Neurol. Psychiatr.* XIV:131–153, 1926.

Penfield WG, Cone W: The acute regressive changes of neuroglia (ameboid glia and acute swelling of oligodendroglia), *Psychol. und Neurol.* XXXIV, 1926.

Perusini G: Grundzüge zur "Tektonik" der weisen Rückenmarkssubstanz, *Journal für Psychologie und Neurologie* XIX:61–89, 1912.

Poldermann H: Die Entdeckung der Mikroglia und ihre Bedeutung für die Neurogliafrage, *Nederlands Tijdschr. v. Geneesk.* LXX(5):537–549, 1926.

Pruijs WM: Über Mikroglia, ihre Herkunft, Funktion und ihr Verhältnis zu anderen Gliaelementen, *Zeitschr. für die ges. Neurol. und Psychiatrie* CVIII:298–331, 1927.

Reynolds FE, Slater JK: A study of the structure and function of the interstitial tissue of the central nervous system, *Edinburgh Med. J.* XXXV:49–57, 1928.

Rezzonico G: Sulla struttura delle fibre nervose del midollo spinale, *Archivio per le Scienze mediche*, 1881.

Río-Hortega P del: Alteraciones del tejido nervioso en los tumores del encéfalo, *La Clínica Castellana*, 1909.

Río-Hortega P del: El tercer elemento de los centros nerviosos I. La microglía en estado normal, *Bol. de la Soc. Esp. de Biol.* IX:68–82, 1919.

Río-Hortega P del: El tercer elemento de los centros nerviosos IV. Poder fagocitario y movilidad de la microglía, *Bol. de la Soc. Esp. de Biol.* IX:154–166, 1919.

Río-Hortega P del: Estudios sobre la neuroglía. La microglía y su transformación en células en bastoncito y cuerpos gránulo-adiposos, *Trab. del Lab. de Inv. Biol.* XVIII:37–82, 1920.

Río-Hortega P del: Histogénesis y evolución normal, éxodo y distribución regional de la microglía, *Mem. de la R. Soc. Esp. de Hist. Nat.* XI:213–268, 1921.

Río-Hortega P del: Estudios sobre neuroglía. La glía de escasas radiaciones (oligodendroglía), *Bol. de la R. Soc. Esp. Hist. Nat.* XXI:63–92, 1921.

Río-Hortega P del: ¿Son homologables la glía de escasas radiaciones y la célula de Schwann? *Bol. de la Soc. Esp. Biol.* X:25–28, 1922.

Río-Hortega P del: Lo que debe entenderse por "tercer elemento" de los centros nerviosos, *Bol. de la Soc. Esp. Biol.* XI:33–35, 1924.

Río-Hortega P del: Condrioma y granulaciones específicas de las células neurógIicas, *Bol. de la Soc. Esp. Hist. Nat.* XXV:34–55, 1925.

Río-Hortega P del: Gliosis subependimaria megalocítica en la senilidad simple y demencial, *Rev. Méd. Barcelona* VIII:85–97, 1927.

Robertson W: A microscopic demonstration of the normal and pathological histology of mesoglia cells, *J. Ment. Sci.* XCVI:724, 1900.

Robertson W: *A text book of pathology in relation to mental diseases*, Edinburgh, 1900, William Clay.

Sánchez M: Sobre la existencia de un aparato especial en los tubos nerviosos de los peces, *Bol. de la R. Soc. Esp. de Hist. Nat.* XVI, 1916.

Sánchez M: El esqueleto protoplásmico o aparato de sostén de la célula de Schwann, *Trab. del Lab. de Inv. Biol.* XIV:253–267, 1916.

Sánchez M: Investigaciones sobre la estructura de los tubos nerviosos de los peces, *Trab. del Museo Nacional de Ciencias Naturales*, serie zoológ. XXVIII:5–96, 1917.

Schaffer K: Über die Hortegasche Mikroglia, *Zeitschsr. für Anatomie und Entwicklungsges.* LXXXI:715–720, 1926.

Schaltenbrand G, Bailey P: Die perivasculläre Piagliamembran des Gehirns, *Journ. fiir Psychol. und Neurol.* XXXV:199–278, 1928.

Schroeder A: Contribución al estudio de la histología normal de los nervios periféricos, *Anales de la Fac. de Med. de Montevideo* X(1):1–18, 1925.

Ségall B: Sur les anneaux intercalaires des tubes nerveux, *Journal de l'Anatomie et de la Physiol.* XXIX:586–603, 1892.

Ségall B: Anneaux intercalaires des tubes nerveux, *Comp. rend. de l'Acad. de Sciences*, 1892.

Spielmeyer W: *Histopathologie des Nervensystems*, Berlin, 1922. J. Springer.

Struwe F: Über die Fettspeicherung der drei Gliaarten, *Zeitschr. für die ges. Neurol. und Psychiatrie* C:450–459, 1926.

Timmer AP: Der Anteil der Mikroglia und Makroglia an Aufbau der senilen Plaques, *Zeitschr. für die ges. Neurol. und Psychiatrie.* XCVIII:43–58, 1925.

Urechia CI, Elekes N: Contribution à l'étude de la microglie, *Arch. intern. de Neurol.* XLV(2):81–96, 1926.

Winkler-Junius F: Die Bedeutung Río del Hortegas Neurogliauntersuchung für die Histopathologie des Zentralnervensystems, *Psychiatr. en Neurol. Bladen* XXVI:91–101, 1926.

Printed in the United States
By Bookmasters

Printed in the United States
By Bookmasters